THE ROUTLEDGE COMPANION TO DANCE IN ASIA AND THE PACIFIC

This Companion documents and celebrates artistic journeys within the framework of rich and complex cultural heritages and traditional dance practices of the Asia-Pacific region. It presents various dance forms from Australia, Cambodia, China, Hong Kong, India, Indonesia, Malaysia, the Philippines, Singapore, Taiwan, Thailand, and the South Pacific. Drawing on extensive research and decades of performative experience as artists, choreographers, producers, teachers, and critics, the authors approach issues of dance and cultural diversity from a theoretical perspective while at the same time exploring change, process, and transformation through dance.

The book discusses themes such as tradition, contemporization, interdisciplinarity, dance education, youth dance, dance networks, curatorial practices, and evolving performative practices of dance companies and independents. It also looks at regional networking, curating dance festivals and spaces that foster collaboration, regional cooperation, and cultural exchange, which are essential features of dance in Asia and the Pacific.

This collection will be of interest to students and researchers of pedagogy, choreography, community dance practice, theatre and performance studies, social and cultural studies, aesthetics, interdisciplinary arts, and more. It will be an invaluable resource for artists and practitioners working in dance schools and communities.

Stephanie Burridge is a research consultant at LASALLE College of the Arts and Singapore Management University. She was Artistic Director of Canberra Dance Theatre (1978–2001) and was awarded the first Choreographic Fellowship at the Australian Choreographic Centre. She is Series Editor for Routledge anthology collections: *Celebrating Dance in Asia and the Pacific* and *Perspectives on Dance, Young People and Change*, co-editor Charlotte Svendler Nielsen, Series Foreword by Sir Ken Robinson. She served as the Research and Documentation Network Co-Chair for WDA Asia Pacific for 15 years and is the recipient of grants for choreography and research from the Australia Council, Arts ACT, NAC Singapore, and Singapore International Foundation.

THE ROUTLEDGE COMPANION TO DANCE IN ASIA AND THE PACIFIC

Platforms for Change

Edited by Stephanie Burridge

Routledge
Taylor & Francis Group
LONDON AND NEW YORK

First published 2022
by Routledge
2 Park Square, Milton Park, Abingdon, Oxon OX14 4RN

and by Routledge
605 Third Avenue, New York, NY 10158

Routledge is an imprint of the Taylor & Francis Group, an informa business

British Library Cataloguing-in-Publication Data
A catalogue record for this book is available from the British Library

Library of Congress Cataloging-in-Publication Data
Names: Burridge, Stephanie, editor.
Title: The Routledge companion to dance in Asia and the Pacific : platforms for change / edited by Stephanie Burridge.
Description: Abingdon, Oxon ; New York, NY : Routledge, 2022. | Includes bibliographical references and index.
Identifiers: LCCN 2021016094 (print) | LCCN 2021016095 (ebook) | ISBN 9780367615079 (hardback) | ISBN 9780367748760 (paperback) | ISBN 9781003160007 (ebook)
Subjects: LCSH: Dance—Asia—Handbooks, manuals, etc. | Dance—Pacific Area—Handbooks, manuals, etc.
Classification: LCC GV1689 .R08 2022 (print) | LCC GV1689 (ebook) | DDC 793.3/1095—dc23
LC record available at https://lccn.loc.gov/2021016094
LC ebook record available at https://lccn.loc.gov/2021016095

ISBN: 978-0-367-61507-9 (hbk)
ISBN: 978-0-367-74876-0 (pbk)
ISBN: 978-1-003-16000-7 (ebk)

DOI: 10.4324/9781003160007

Typeset in Bembo
by Apex CoVantage, LLC

CONTENTS

FIGURES

CONTRIBUTORS

Jala Adolphus is an independent producer based between Australia, Indonesia, and Europe. She works with a number of leading Indonesian choreographers and directors, supporting the long-term map of projects from inception through to production and touring circuits. She currently produces the work of film and stage director Garin Nugroho and choreographer Rianto. She has worked on Indonesian programming and was Editor of Tanzconnexions, the Goethe Institut platform for contemporary dance in the Asia Pacific (2014–2017). She has been a member of the Asian Producers Platform since 2014 and is a core member of the Asia Network for Dance AND+.

Priyanka Basu (PhD) is the Project Curator of 'Two Centuries of Indian Print' at the British Library and a visiting scholar at the SOAS South Asia Institute in London, UK. She received her PhD from SOAS, completed on a Felix scholarship. She is currently finishing her manuscript, *The Cultural Politics of 'Folk': Perspectives from India and Bangladesh*. Her research interests include theatre, performance and film histories, book history, gender, and dance studies. She curates and runs the British Library South Asia Seminar series. She is trained in the Indian classical dance form of *Odissi*.

Vanini Belarmino is currently Assistant Director (Programmes) at National Gallery Singapore. In a career that has spanned more than 25 years, Vanini has conceived and spearheaded artistic programmes, collaborating with more than 1,000 artists across disciplines in Asia, Europe, and the Middle East. The resulting works create new experiences by initiating a series of encounters between artists and the audience within the public space, such as *I'mPULSE, Pointe to Point, ZENSORS, Intimate Moments*, and the inaugural year of ArtScience Late for ArtScience Museum. A recipient of the Asian Cultural Council Fellowship in New York, Vanini has worked with the Cultural Center of the Philippines, Ballet Philippines, European Cultural Foundation, and Asia-Europe Foundation and has served as General Manager for dance.MNL, the Philippine Dance Festival, and the first Singapore International Jazz Festival.

Anna C.Y. Chan is the Dean (Dance) at The Hong Kong Academy for Performing Arts and an experienced producer, curator, and arts consultant. She was the Vice-President (East Asia) of the World Dance Alliance Asia Pacific, the Chairperson of the Hong Kong Dance Alliance, and the inaugural Head of Dance at the West Kowloon Cultural District and is currently a committee member of the Tai Kwun Culture and Arts Centre, advising their curatorial policy and programming in visual art, performing arts, and heritage areas. She has launched many international networking projects, including Producers Network Meeting and Forum, and founded the Asia Network for Dance AND+.

Chey Chankethya started her classical dance training at the age of six and obtained her BA in choreographic arts from the Royal University of Fine Arts in 2005. She also holds a BA in English from the Royal University of Phnom Penh and a master's degree in dance/choreography from UCLA as a Fulbright Scholar, supported in part by the Asian Cultural Council in 2014. She has created a large body of work in both classical and contemporary forms that have toured internationally, including the acclaimed solo *My Mothers and I* (2015). She is a 2013 Mo Ostin Performing Arts Award winner and a Georgetown University Fellow for 2017–19 at the Lab for Global Performance and Politics. She is now teaching at the Royal University of Fine Arts, Faculty of Choreographic Arts, Phnom Penh.

Chen Ya-Ping (PhD) is Associate Professor at the Graduate Institute of Dance, Taipei National University of the Arts, Taiwan, and the author of the Chinese monograph *Enquiry into Subjectivity: Modernity, History, Taiwan Contemporary Dance* (2011). Her research articles have been published in English, Chinese, and Japanese in various academic journals and dance studies–related anthologies. Her research interests include modernity and dance history; theories of corporeality, place, and memory in relation to dance studies; and dance criticism. She is currently the President of the Taiwan Dance Research Society (TDRS) and was the Chairperson of the Graduate Institute of Dance at TNUA from 2015 to 2018.

Angela Conquet is a dance specialist and curator based in Australia. She is the Editor of *Dancehouse Diary*, an accessible publication that connects the body to the body social and the body political. Her 20-year curatorial and managerial career spans two continents; a multiplicity of disciplines, contexts, and networks; and an unfaltering commitment to connecting dance artists' ideas to new contexts and new audiences. She was Artistic Director/CEO of Dancehouse Melbourne from 2011 to 2020 and Director of Dance at Mains d'Oeuvres, Paris, from 2006 to 2011. She is a member of the Asia Network for Dance AND+.

Pornrat Damrhung (PhD) is Professor in the Department of Dramatic Arts, Faculty of Arts; a research professor in the Division of Research and Innovation at Chulalongkorn University; and an associate member of The Royal Society of Thailand in the Academy of Arts. In 2018, she helped found the first research cluster at Chulalongkorn University in the humanities, centred on innovation in the arts and culture, and runs the section on 'Cultural Ecologies of Performance: Creativity, Research and Innovation'. She recently completed a senior research scholar grant (2016–2019) from the Thailand Research Fund to nourish young researchers and artist networks around Thailand. For more than 30 years, her research and stage work have sought to make traditional Southeast Asian performers, stories, dances, and music part of contemporary performances able to move today's audiences.

Peter Gn is currently a PhD candidate at the Victorian College of the Arts, University of Melbourne (2019). His versatility in the dance field in Singapore has seen him fulfil roles as dancer, dance educator, and choreographer through the years. He has an MA from Roehampton University (London) and, as a recipient of the Trinity Laban Award, trained as a choreographer at the Trinity Laban Conservatoire of Music and Dance (London). His current work in Singapore includes providing national-level support and consultancy to schools and arts institutions. Peter also holds a master's (by research) in philosophy from Murdoch University (Australia) with a master's in metaphysics.

Bilqis Hijjas writes about dance and produces, choreographs, and performs in Malaysia. While studying at Harvard University for her BA in social studies, she worked with a number of campus dance companies. In 2002, Bilqis completed her graduate diploma in choreography at the Victorian College of the Arts, Melbourne, and began her formal dance career with the Melbourne Fringe Festival in 2005. In 2006, Bilqis moved home to Malaysia to establish the amateur contemporary ballet group Balletbase. Bilqis also directs the dance programme at private arts centre Rimbun Dahan, which offers residencies for contemporary dance choreographers. She has a master's degree in applied anthropology and international development and is currently the Secretary of MyDance Alliance.

Kathy Hong is the Chairperson of the Performing Arts Network Development Association, an association of arts managers in Taiwan, and a consultant working within the theatre, visual arts, and event industries. She is an arts professional with extensive experience in theatre and international collaboration, including producing festivals and branding arts organizations. She was Executive Director of OISTAT – International Organization of Scenographers, Theater Architects and Technicians, 2012–2015, and Marketing and Communications Director of National Kaohsiung Center for the Arts, 2015–2019. She is a core member of the Asia Network for Dance AND+.

Ophelia Jiadai Huang is a cultural manager, dramaturg, and performance maker based in Shanghai, working currently as Director of International Projects at the Shanghai Dramatic Arts Center and Programme Director of ACT Shanghai International Theatre Festival. With experience working in an international cultural agency in China for more than a decade, her research interests also include cultural policy and diplomacy. Ophelia holds MAs in performance research from the University of Warwick and interdisciplinary arts theory from University of Arts in Belgrade. She is a core member of the Asia Network for Dance AND+.

Sarah Knox is a lecturer and doctoral candidate in dance studies at the University of Auckland, New Zealand. She has had an extensive career as a performer, choreographer, and teacher. Her research interests include collaboration, dance education, choreography, and creativity. Sarah was the co-author for the Artists' Voices chapter of the book *Moving Oceans: Celebrating Dance in the South Pacific* (2014), along with Dr Rose Martin.

Lin Yatin (PhD) is Assistant Professor at the School of Dance, Taipei National University of the Arts. She received her PhD in dance history and theory from the University of California,

Riverside, researching on Cloud Gate Dance Theatre. She currently leads a project funded by the National Science Council analyzing dance festivals and cultural tourism in the Asia-Pacific region. Former Secretary General of the Dance Research Society Taiwan, she serves on the Board of Directors of the Society of Dance History Scholars (USA).

Rose Martin (PhD) is Associate Professor of Arts Education with a focus on multiculturalism at the Norwegian University of Science and Technology. She also holds positions as International Research Advisor for Nanyang Academy of Fine Arts, Singapore, and Adjunct Professor at Northwest Normal University, China. Rose has extensive experience in research and teaching around the world, publishing books, book chapters, and journal articles focused on post-colonial dance education, dance ethnography, and arts and politics. Rose's research had a strong focus on the regions of the Middle East, China, and Scandinavia.

Marilyn Miller began dancing at the age of five, taking formal classes in classical ballet whilst learning Aboriginal dance around a campfire at the beach. As a teenager, she studied at the Australian Academy of Ballet and gained a place at NAISDA Dance College. She went on to be the sole female co-founder of Australia's first Indigenous contemporary dance company, Aboriginal Islander Dance Theatre – the Company (or AIDT – the Company). After a successful career as a dancer and choreographer, Miller moved into arts management. She has held numerous high-profile positions in the arts, including Chair of the NSW Ministry of the Arts Dance Committee and co-Deputy Chair of Ausdance National. She established BlakDance Australia and is studying for an MFA in cultural leadership at Australia's National Institute of Dramatic Arts (NIDA).

Mohd Anis Md Nor (PhD) is a retired professor of ethnochoreology and ethnomusicology at the University of Malaya and currently an adjunct professor at Sunway University and Managing Director of Nusantara Performing Arts Research Centre in Kuala Lumpur. He has pioneered the study of *zapin* dance and music in Southeast Asia and has authored more than 20 books, 93 chapters/articles, and 151 keynotes/conference papers on the performing arts of Malaysia and Southeast Asia, focusing on music, dance, and theatre, and has enjoyed tenure as a visiting professor at prestigious universities in the United States, Europe, and Asia.

Hun Pen began dancing when she was five years old. In 2002–2003, she studied at Wesleyan University in Connecticut, USA, and in 2003, she obtained a BA in choreography from the Royal University of Fine Arts. She has worked with Cambodian Living Arts as a co-director for the Mohour Srop project and Amirita Performing Arts as an assistant project coordinator. She studied modern dance and arts management at the Cite International des Arts in Paris and in 2005 worked with the Gallota Dance Company in France as a cultural ambassador, choreographer, and dancer. She has a master's degree in Southeast Asian studies from Chulalongkorn University, Thailand. As a cultural leader, she serves as a deputy director of the Department of Performing Arts, where she leads, teaches, and directs projects, including choreographing, to numerous students and artists.

Melissa Quek, prior to graduating with honours from New York University Tisch School of the Arts, was a dance-artist performing professionally with the tammy l. wong Dance Company in Singapore. With the National Arts Council Arts Bursary (Overseas), she attained a

BFA with a double major in dance and English and American literature. In 2011, she completed her MA in dance cultures at the University of Surrey. She continues to perform and choreograph and write critiques for local newspapers – she is Head, School of Dance and Theatre, at LASALLE College of the Arts, Singapore.

Michael H.B. Raditya is a researcher, critic, and writer who lives in Yogyakarta, Indonesia. He graduated from Universitas Gadjah Mada, Indonesia, with a major in anthropology (BA) and performing arts and visual arts studies (MA). He is an editor for *Journal Kajian Seni* and has authored Indonesian books on performing arts, including *Life for Dance* (2016), *The Power of Art* (2017), and others. He contributes to several online publications around the Asia Pacific region.

Urmimala Sarkar (PhD) is an associate professor of Theatre and Performance Studies at the School of Arts and Aesthetics at Jawaharlal Nehru University, New Delhi, India. She is the President of the World Dance Alliance – Asia Pacific and the Co-Chair of the WDAAP Research and Documentation Network Her current research focuses on marginalization and living traditions, politics of performance, gender and dance, and performance as research (PaR). She is the Co-Editor of the peer-reviewed web journal *JEDS* (*Journal for Emerging Dance Research*).

Lowell Skar is a lecturer in the Faculty of Arts at Chulalongkorn University in the BALAC Program focused on global cultures and media. His diverse interests and long experience in East and Southeast Asia centre on understanding how modern cultural life in the region, including the performing arts, is bound up in our involvement with contemporary transnational practices and imaginaries. He has taught a wide variety of courses in cultural theory, cultural history, transnational cultural life, and science and technology studies for undergraduate and graduate students and has guided student theses.

Jill Sykes (AM) began reviewing dance in London and is Dance Critic for *The Sydney Morning Herald.* She has been a freelance arts journalist most of her career, writing about theatre, music, and the visual arts as well as dance. For 13 years, she was Editor of *Look*, the membership magazine of the Art Gallery Society of NSW, and is the author of the book *Sydney Opera House – From The Outside In* and editor of the book of the TV series *Wine Lovers' Guide to Australia.* She was awarded a Lifetime Achievement Award for contributions to dance in 2020 at the Australian Dance Awards.

Faith Tan has over 17 years of programming, producing, and management experience in the performing arts. She is currently the Head of Programme Development at Dance House Helsinki, Finland. She is in charge of the house's programming, including the opening season and the commission of new work. Formerly, she was the Head of Dance and Theatre in Singapore's national arts centre, Esplanade – Theatres on the Bay. Faith led *da:ns* – the largest international platform for dance within Southeast Asia. Faith is a founding member of the Asia Network for Dance AND+. She holds a master's degree in arts and cultural management.

Emily Wilcox (PhD) is Associate Professor of Modern Chinese Studies in the Department of Asian Languages and Cultures at the University of Michigan, USA. She is author of

Revolutionary Bodies: Chinese Dance and the Socialist Legacy (2019), winner of the Dance Studies Association de la Torre Bueno Prize. Wilcox is co-editor of *Corporeal Politics: Dancing East Asia* (2020), co-creator of the digital archive *Pioneers of Chinese Dance*, and co-curator of the 2017 exhibition *Chinese Dance: National Movements in a Revolutionary Age, 1945–1965*. Wilcox has published over 20 academic articles and essays on Asian dance and performance.

ACKNOWLEDGEMENTS

The majority of chapters in this book are drawn from the original Routledge *Celebrating Dance in Asia and the Pacific* series. This ongoing endeavour is supported by the extensive membership networks of the World Dance Alliance of Asia and the Pacific (WDA), and I warmly thank the WDA for enabling these connections. Selected chapters that first appeared in the Routledge *Celebrating Dance in Asia and the Pacific* series have been modified and updated – these are central to this book, including Australia, Cambodia, India, Malaysia, Taiwan, Singapore, and the South Pacific. I deeply respect and thank the contributions of the co-editors of these books – Fred Frumberg, Urmimala Sarka Munsi, Mohd Anis Md Nor, Yunyu Wang, Julie Dyson, Caren Carino, Ralph Buck, and Nicholas Rowe.

The scope for this volume has widened to include chapters from China, Hong Kong, Indonesia, the Philippines, and Thailand. The objective of the series is to present the views of eminent scholars, journalists, and commentators alongside the voices of a new generation of choreographers working from tradition to create new forms of expression in contemporary dance. I would like to acknowledge all of the outstanding contributors; they reflect the diversity and connectivity of the dance sector with views and insights shared by performing artists, choreographers, producers, teaching artists, critics, curators, researchers, and university faculty. Last, I sincerely thank Routledge, India, for their foresight in commissioning this volume and their continued support for dance in the region.

INTRODUCTION

Stephanie Burridge

The Routledge Companion to Dance in Asia and the Pacific: Platforms for Change documents and celebrates artistic journeys that exist within the framework of rich and complex cultural heritages and traditional dance practices. Selected chapters that first appeared in the Routledge *Celebrating Dance in Asia and the Pacific* series have been modified and updated – these are central to this book, including Australia, Cambodia, India, Malaysia, Taiwan, Singapore, and the South Pacific. The scope is widened to include accounts of dance from China, Hong Kong, Indonesia, the Philippines, and Thailand. A special feature of the volume is the embodied experiences of the writers – almost all are from the countries they write about, and they bring decades of expertise in research and creative practice. The authors share multiple perspectives and insights from their roles as performing artists, choreographers, producers, teaching artists, critics, curators, researchers, and university faculty.

Watching a dance performance in the Asia Pacific region involves a complex of threads and pathways. Artists are incorporating a diversity of movement languages, dance philosophies, techniques, and narratives that interweave in divergent and interconnected lines around the region. Choreographers work through their embodied cultural memories and multiple dance traditions – these often co-exist and merge with western contemporary dance forms. Dance around the region is moving rapidly with a creative confidence that is stimulating audiences and re-framing the genre of contemporary dance in the region and beyond. The challenge is to facilitate dance opportunities that link histories and communal cultural knowledge with new creative explorations led by individual artists.

Historically dance has had an important function as a continuum that expresses and celebrates the rituals of life and death and contact with the spiritual world. The traditional role of dance in many Asian countries was to perform for the royal family or the gods. Facing forward without turning the back and dancing in a lateral spatial plane is an element in all court dancing; stages were generally small, and the intricate, heavy costumes and weight of the headdresses restricted the movement through space. The purpose of the dance was to reinforce the prevailing order of life and affirm the traditions of the culture – to act as a link from the present to the past. The intention is deeply respectful. Indigenous and folk dances are also part of this mix, and in some countries like Australia, Indonesia, Philippines, and

DOI: 10.4324/9781003160007-1

Taiwan, these influences have been fundamental to developing unique dance vocabularies and choreography. In many countries of the region, the recognition of movement forms with deep internalization of mind/body/spirit connections are essential to dance training; Indian yoga practice, Indonesian *silat* martial arts, and Chinese *qi gong* are examples. Most traditional dance forms exist in a multi-disciplinary manner – for example, it would be inconceivable to practice the intricacies of Bharatanatyam, one of the eight classical Indian dance forms, without live musical accompaniment.

Pedagogy

Intense dance training underpins explorations into contemporary dance anywhere in the world. The body is tuned, prepared, and disciplined to create lucid movement plus having the facility and stamina to rehearse. In performance, the dancer needs to be physically articulate to define the intention of the choreography and have the virtuosity and vitality to give energy to a performance. A guru, or dance master, is deeply revered in Asia, and young dancers in many countries of the region learn traditional dance through this rigorous approach. Hours of practice over many years are needed to develop the strength and flexibility to achieve the complex postures and gestures of classical dance. This requires stringent disciplining of the body, mind, and spirit. Only a handful of dancers achieve perfection in the classical forms – each has endured the sacrifices that becoming a professional entails. Thus, for many dancers from the Asia Pacific region, working with tradition is a constant challenge. Respect for the form and the master teachers are fundamental and learning the vocabulary and nuances of each dance time consuming. For dancers who want to experiment and push some boundaries, going outside their cultural modality and journeying towards contemporary trajectories may be challenging. Methods that are commonly used to inspire creativity in the West are uncommon in traditional practice in many places. These include the 'workshop' process, contact improvisation and collaborative exploration and creativity. Influences filter through as dancers have the opportunity to attend master classes, choreographic workshops, or dance college courses elsewhere or increasingly in their home countries as universities and conservatories for full-time dance training open up. In the Asia Pacific, insightful selection of visiting teachers and choreographers, typically by companies and festivals, has achieved some exciting cross-cultural collaborations; however, the polarity is the rejection of Western or outside influences and a re-examination of local culture and the rich eclecticism of classical, folk, and indigenous dance languages. This approach has increasingly gained traction – in this volume, these opposing positions come together to various degrees across the span of the countries discussed.

Although the class situation in the West may seem the same for some dancers, by comparison to the guru or master teacher system, there is a spectrum of training options for Western dance students to choose from. Tertiary training academies are globalized and offer courses that include ballet and contemporary dance plus myriad other options such as urban street dance, jazz, ethnic forms, salsa, and Broadway. Improvisation and composition are also the norm along with theory subjects such as aesthetics, criticism, and dance history. This diverse mix reinforces the notion of the self as the centre of the training and is a system where individual skills are encouraged and the dancer prepared for diverse dance industry options both locally and internationally. Similarly, training that combines Eastern and Western genres with varying emphasis exists in many Asian countries through tertiary dance programmes.

Cultural exchange, commissions, and curating regional festivals

The development of new Asian movement vocabularies arises from exploring within the traditions and philosophies that the region encompasses and combining these connections with personal concerns and larger themes about the society and state of the world. Like all dance makers, Asian choreographers search to find a unique voice in concept and movement vocabulary – creativity, virtuosity, and perseverance have been the foundation for both emerging and the previous generations of contemporary Asian dance makers.

Over the past couple of decades, there has been an explosion of creativity and an evolution in the generic forms of dance, and cultural exchanges remain an important aspect of the evolving Asia Pacific dance ecology. These connections have resulted in a network of festivals that feature regional artists from Asian countries such as China, Japan, Taiwan, Korea, Thailand, Cambodia, India, Malaysia, Hong Kong, Indonesia, and further afield to include Australia, Europe, and the United States. Festival performances, showcases of works in progress, and after-show dialogues are part of almost every independent choreographer's and each local dance company's annual schedule. There appears to be an established geographical 'creative' zone where regional artists freely mix, borrow, amalgamate, and blend forms that redefine dance into an evolving umbrella known as 'Asian contemporary'. Embodied cultural knowledge, and confidence in rich traditional dance heritages, underpin bold innovations by regional choreographers These initiatives have encouraged a thriving group of independent dance artists and small companies who are commissioned by dance festival curators to make new work. Recently, formalized networks between regional centres have evolved and become essential players in enabling talented regional artists to tour and be seen outside of their country. Dance curators, festival directors, dramaturgs, producers, and independent dance centres are central to this story of the 21st-century dance ecology in the region. Chapters on Hong Kong, the Philippines, and Indonesia in this book specifically refer to some of this activity.

Digital connections and collaborations are integral to this scene, with numerous performative projects that are a hybrid mix of the virtual space and live actualizations. This volume recounts some work by innovative dance film-makers. During COVID-19, digital platforms have exploded with creative, and necessary, responses by dance artists to keep their work alive and offer audiences a plethora of responses, from personal sharing of their homes as sites for dance making to large established companies offering free live streaming or posting past productions online.

Conclusion

The new dance of this millennium is grounded in multiple cultural identities, and it is these complexities that throw out new and often perplexing questions to audiences. Controversial themes, unrecognizable movement, conceptual philosophically based practice, the unfamiliar . . . all of these occur in a contemporary dance performance. It is an art form that has been pushing boundaries and stretching imaginations for over a century. The need to explore, express, surprise, and subvert the conventional norms of society are intrinsic to the genre. Added to this narrative is the impact of technology – dance in the digital age has changed creative practice, performances, marketing, documentation, and archiving. Everyone can film a dance in their living room and post it for millions to see – bloggers and influencers effecting instant pathways to trends with dance is an integral part. Edges become blurred as audiences navigate through countless posts of one-off clips and dance mashups on social media appearing alongside choreography from world-class professional companies and innovative dance independents. Dance education, dance history, cultural contexts, and skills in critical analysis are important partners for the next generation of dance artists and their audiences.

At the time of writing, the world is in the grip of the COVID-19 pandemic. This has profoundly changed dance for many performers, choreographers, educators, researchers, participants, and audiences. At this point, we cannot imagine the long-term impact and changes to the dance ecology. Immediate concerns are grappling with everyday issues from loss of income, cancellations, indefinitely postponed creative projects, touring, and cultural exchanges. Many are forced to innovate at their own expense . . . scrambling to upload teaching materials that engage students in digital dance classes and lectures in the online space instead of the dance studio and posting performances and choreography. How can dancers maintain professional-level training and consider their emotional well-being amid the realities of confinement and constriction? For dancers who typically engulf space and seek freedom of expression, this is extremely challenging. Survival is paramount . . . sustainability necessary. Some countries in the region have supported the arts through various grants and subsidies, but for many there is little hope and a daunting task ahead to rebuild careers and keep passion alive until we emerge from this crisis. Inevitably there remain many tensions, dilemmas, and uncertainties for both the artists and the audiences as familiar ground is shifting. Dancers in Asia and the Pacific will embody and make meaning from these rapid changes and the shock and trauma of this crisis. Yet dance has survived for millennia. It will continue to not only adapt but surge ahead with vigour, imagination, and new stories to tell.

Australia

1

SHAPING THE LANDSCAPE

Jill Sykes

It may be a 'new' country in terms of European settlement, but Australia is believed to have the oldest form of continuously practised dance. It goes back at least 65,000 years as an integral part of the Indigenous culture and is still being passed on to new generations. Most excitingly, it has also become the basis of contemporary styles for which the leading choreographers follow guidelines from the Indigenous elders as they strike out in fresh directions. This is the only form of dance, old and new, that Australia can claim as its own. It has had a thrilling rise to mainstream appreciation in the past 45 years or so and is considered so important that four contributors to this book have written about different aspects of its development.

Aboriginal dance had, of course, been noted by dance practitioners in the theatre before that. But inevitably the creative team and performers were European, and we look back on their well-meaning attempts with a shudder: in the 1950s, blacking up and all-over black costumes were acceptable in representing people with dark skins.

Times have changed, and Australia now has highly trained dancers from Aboriginal and Torres Strait Islander backgrounds who have a breadth of knowledge and experience that they can apply to the sophisticated demands of a company like Bangarra Dance Theatre. This company celebrated an important anniversary in 2019 with a triple bill under the self-explanatory title *30 Years of Sixty Five Thousand*. One of those works was Jiri Kylian's *Stamping Ground*, inspired by an Aboriginal gathering the choreographer attended in 1980 and coming full circle in a powerfully danced presentation by Bangarra's Indigenous performers in Kylian's characteristic European style. Australia's Indigenous dancers perform on mainstages around the country and overseas, as well as finding their way to new audiences via new media. That cheeky version of Zorba's Dance by the Chooky Dancers on YouTube was enjoyed by millions of people around the world and helped these Elcho Island performers from the Northern Territory to a commissioned work that was presented in 2010 at the Adelaide Festival and the Sydney Opera House. Under their formal language title, *Djuki Mala* – you can see where the Chooky came from – they are still touring their mix of Indigenous and western movement, beloved by their audiences.

This broader exposure to forms of Indigenous performance – its humour as well as its serious dance and music from the past and present on pertinent, often social themes – has

DOI: 10.4324/9781003160007-3

been reinforced by a parallel interest in Aboriginal art. As well as providing entertainment and stimulating ideas, it has been a bridge to discovering more about their own country for many Australians.

With the exhilaration of developing a uniquely Australian dance, the traditions of dance as a theatre artform have not been forgotten. In fact, ballet is the mainstay of this country's dance audience if you take the biggest, best-funded company with the largest following as your guide – and in that, it is probably no different from other western nations. But in the third decade of the 21st century, Australia has many other styles of dance with robust support and practitioners. Most of them are covered in detail in this book. It is a vibrant dance scene with the promise of many organizations around the country getting young people to dance – not only encouraging them through performances that appeal to them but getting them to take part.

Australia's isolation, both on the globe and the distance between its population centres, turns out to be something of an advantage to creativity. It is not like some national dance scenes in Europe where a popular style produces clones. In this large continent, there are dozens of small groups responding to their communities and doing their own thing. Though you can spot some shared influences, you can't pin down a wave of similarities. Australian dance is richly endowed with individuals.

Small groups and independents are unshakeable in their devotion to dance and find ways of sharing it with others, no matter the financial pressures where funding favours the old over the new. The larger state and federally funded companies are constantly drawing on familiar stories or making new versions of tested favourites in a bid to hold onto their audiences.

Commercial dance does well in Australia, with its profusion of musicals. Multicultural dance survives and sometimes thrives. Dance theatre and physical theatre are increasing in practice and popularity. Opportunities for disabled dancers are growing. Postmodern dance has all but disappeared. Hip-hop has its own following and is permeating performances of all kinds.

At the time of writing, six months into COVID-19, it is impossible to predict the future of dance companies, with venues closed and social distancing in place. Many have been reaching out digitally to their audiences by offering free or modestly priced classes online and beaming out recordings of past productions plus new solos and duets as restrictions allow. The public has responded by giving generously to help the dancers keep going.

But to return to Australia's biggest flagship company – The Australian Ballet. Although it builds its subscription audience on popular classics – it has two versions of *The Nutcracker* in its repertoire, one traditional, the other with a more contemporary Australian slant by Graeme Murphy – it has always mingled old and new since its foundation in 1962.

In 2020, the farewell year for David McAllister after 20 years as artistic director, it was due to present a fairly typical cross-section of works by Australian and overseas choreographers. *The Happy Prince*, a full-length work by Graeme Murphy, just managed its world premiere before COVID-19 shut down the venues. Audiences were to have been treated to Yuri Possokhov's *Anna Karenina*, Alexei Ratmansky's revival of Marius Petipa's *Harlequinade*, a new short work by talented company member Alice Topp, established short works by Australians Stephen Baynes and Tim Harbour, two from the repertoire by Britain's Wayne McGregor, and Frederick Ashton's classic *A Month in the Country*. David Hallberg has been appointed the Australian Ballet's artistic director from 2021. He is best known as a principal dancer with American Ballet Theatre, but he has also danced as a member of the Bolshoi Ballet and been a guest artist with Britain's Royal Ballet and the Australian Ballet, for which he has made regular appearances over the past decade. He has danced the great classical roles, worked with today's choreographers, and directed an ABT workshop for emerging choreographic talent, all of which suggests a good background for leading the Australian Ballet.

The West Australian Ballet is the country's oldest company, founded in 1952 by Madame Kira Bousloff, a former member of the Ballets Russes. Its current artistic director, Aurelien Scannella, was born in Belgium, and he has danced classical and contemporary works. Since his appointment in 2013, he has called on local and overseas choreographers to build on the company's repertoire of classics and new works on popular themes, including *Giselle, Cinderella, The Great Gatsby*, and Krzysztof Pastor's *Dracula*.

The Queensland Ballet has moved ahead with a burst of energy and achievements since Li Cunxin took over as artistic director in 2013. This Chinese-born dancer, whose early career was detailed in the book and film *Mao's Last Dancer*, is providing skilful leadership that is developing the company's character and engaging audiences. An infusion of Cuban dancers has added to the strong local performing talent in a blend of classical and contemporary ballets such as *The Sleeping Beauty, Swan Lake, A Midsummer Night's Dream, Peter Pan*, and *Carmen*. Disappointed by COVID-19's demolition of its 60th anniversary plans in 2020, the company opened up a free digital season of brief works made and performed by its dancers.

After an extraordinarily creative period of 31 years under the leadership of Graeme Murphy, the most interesting and imaginative choreographer Australia has produced, the Sydney Dance Company (SDC) appointed Rafael Bonachela as artist director in 2009. Spanish-born Bonachela's varied background working in London – Rambert Dance Company, choreography for pop singer Kylie Minogue, his own company – has brought fresh vitality to the SDC with new dancers and a range of guest choreographers such as Emanuel Gat, Jacopo Godani, Alexander Ekman, and the promising young Australian Gabrielle Nankivell to complement his own energetic works on thoughtful themes. The company has a busy schedule in Australia and regular tours overseas.

In Adelaide, the Australian Dance Theatre had a similar spurt of action when Garry Stewart took over as artistic director in 2000. He has maintained his energy in making slow-release works of action-packed velocity that have been seen around the world. He has had serious fascinations with technology – including intriguing collaborations with New York photographer Lois Greenfield and Canadian robotics artist Louis-Philippe Demers – but was more concerned with the natural world in his recent work.

Also in South Australia, Leigh Warren retired from the small company that carried his name after 23 years and many successes, including the direction and choreography of a trilogy of Philip Glass works in partnership with the State Opera South Australia. His company was remade in 2019 as Dance Hub SA to nurture independent dance in that state.

Australian audiences have always relished theatrical dance. The country's collective love affair with the Ballets Russes, starting at the end of the 1930s, set a benchmark that has been maintained, even by the balletomanes. And for everyone else, it is getting to the point where you might wonder where the theatre ends and the dance begins and vice versa.

Lucy Guerin has approached some topical and surprising subjects for her eponymous dance company, including the tragic collapse of a bridge being built in Melbourne nearly

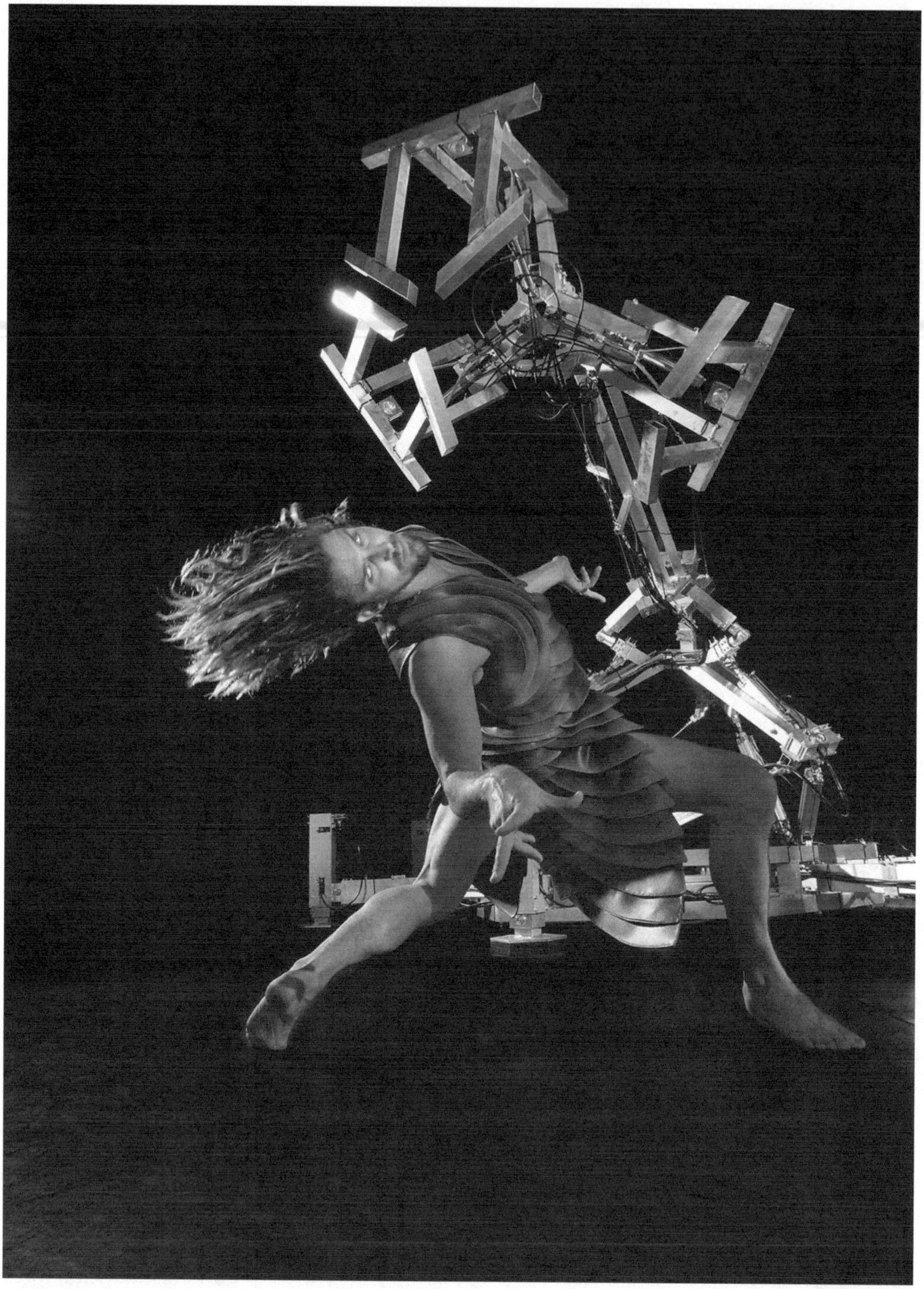

FIGURE 1.1 Australian Dance Theatre, *Devolution* (2006)

Source: Dancer: Tim Ohl

Photograph: Chris Herzfeld

Source: Supplied by Australian Dance Theatre, in the collection of Ausdance National

30 years back and an amusing and insightful exploration of dance training in *Untrained*. Born in Adelaide but established in Melbourne with Lucy Guerin Inc since 2002, she has won many awards in Australia and a Bessie in New York with choreography that pushes the boundaries of dance thought and action. Her recent works include *The Dark Chorus, Attractor*, and *Make Your Own World*. Stephanie Lake is also based in Melbourne. Her small company was formed in 2014, but she often creates work for much larger groups, producing some of the most arresting choreography to emerge in Australia over the past decade, feisty and intelligent. A multi-award winner, she has taken her work overseas, including Paris, Dublin, Singapore, and Hong Kong. Melbourne also has its small groups that dance on a classical base. Melbourne Ballet Company, founded in 2007, has made strong community connections through tours in Australia and New Zealand. Victorian State Ballet, established in 2003, mostly tours regional Victoria.

Kate Champion founded her dance theatre company Force Majeure in Sydney with great success for her treatment of unlikely themes, weaving stories out of research and workshops in an organic development that was matched by the naturalistic style of movement. She has now moved to freelance theatre direction, and Force Majeure continues, devising works with multi-disciplinary artists under the leadership of Danielle Micich.

Tess de Quincey, based in Sydney but performing nationally and internationally with the company that bears her name, has built a following for her 'body weather' style. It grew out of Japan's butoh and expanded in workshops undertaken in Australia's parched interior to bring out contrasting aspects of performance that are then intensified down to their engrossing essence for presentation in venues that are often out of the ordinary.

Expressions Dance Company, founded in Brisbane by Maggi Sietsma in 1984, has a new artistic director, Amy Hollingsworth, and a new name: Australasian Dance Collective. At the time of writing, it was hoping to make its first appearance under its new title at an outdoor performance of *Arc* for the Brisbane Festival, COVID-19 permitting.

Also in Queensland, Dancenorth is based in the far northern city of Townsville. This small company, so distant from any of its peers, has been going since 1985 as a professional contemporary dance company, initially directed by Cheryl Stock, and is currently under the direction of Kyle Page.

Legs on the Wall has created a world of its own. Its starting point in 1984 was acrobatic prowess as much as anything, but the group turned this ability into a tool for theatre that engrossed, astonished, and moved audiences. At the 2000 Sydney Olympic Games, they scaled a tall building and told a story at the same time, winning new fans in the streets below. The group's influence, through the involvement of current and former performers, has reached out into straight theatre and even opera in a collaboration with Opera Australia. Its most recent success, *Man with The Iron Neck*, was due to tour in 2020.

Stalker has a more circus-inclined approach to an idiosyncratic style of physical theatre that ranges through contemporary dance and martial arts, powered by a strong philosophical stance in its commitment to inclusion and reconciliation. It is often seen in unusual natural settings that enhance its individuality, and it is still making an impact under one of its original founders in 1985, David Clarkson.

For two decades, Stalker had an offshoot group, Marrugeku, which in 2016 became independent and has since gone from strength to strength in intercultural performance under the leadership of Dalisa Pigram and Rachael Swain. Working between Sydney and

Broome in northern Western Australia, Marrugeku has distinctive Indigenous and Asian strands.

And then there is the magical group Gravity & Other Myths that sprang into vivid existence in Adelaide in 2009. It celebrates ensemble virtuosity in acrobatics and physical theatre with pieces such as *Backbone* and *Out of Chaos*, which were hits of the Adelaide Festival in 2017 and 2019, respectively, before they toured.

For something completely different, BalletLab works out of Melbourne under the direction of Phillip Adams, creating pieces that range from dance to installation, and do well overseas. In Adelaide, Restless Dance Theatre integrates young performers with and without a disability. It had notable success at the 2019 Adelaide Festival with *Zizanie*, choreographed by Meryl Tankard.

Increasing numbers of Australian choreographers are putting dance on screens, large and small. Some, like Sue Healey, work alone and in conjunction with stage presentations. She has made a strong impression with her portraits series, combining filmed and live action of individual performers.

On the subject of technology, there was one Australian choreographer obsessed by it for his stage presentations. In between works that make something special out of deeply ordinary themes – *I want to dance better at parties* – Gideon Obarzanek explored interactive video technologies with Frieder Weiss, using movement as a trigger for lighting changes, first in the solo *Glow*, then as an ensemble *Mortal Engine* for his company Chunky Move. Obarzanek has since moved on to freelance directing of large events, and Chunky Move has been led since April 2019 by Antony Hamilton, Kristy Ayre, and Freya Watson, 'aiming to increase the visibility of contemporary dance as an everyday artform'.

The sheer size of Australia may be an advantage creatively, as already mentioned, but the cost of touring in a country close to the size of the United States, but where population centres are so far apart, is prohibitive. Which means dancegoers don't have much opportunity to see what is happening elsewhere. Multi-arts festivals help, though they are more likely to bring in foreign groups, with Dance Massive in Melbourne the standout survivor of the artform. And once a year, the Australian Dance Awards bring together dancers from different parts of the nation to stage an awards programme that makes as much a contribution to a national mix as it can.

So you really have to go to Darwin to see what Tracks is doing – which is quite unlike any other dance company in that it is so closely moulded to the fascinating multicultural mix of that city on the country's northern edge: facing Asia, with the outback and Australia's Indigenous people all around it. The diversity of Tracks is unique, from funky youth dancers to 'showgirl grannies', as they like to call them, and ochre-painted elders – sometimes separately, sometimes together.

Tasmania's Tasdance, almost as far south as you can travel in Australia, is another company that goes its own way – which it did with distinction under the direction of Annie Greig from 1997 to 2015 – and it continues to lead a determined existence. Location has important bearing on the way people make dance. In Perth, for instance, it might be something to do with that city's isolation from the other Australian capitals, or the presence of WAAPA, the Western Australian Academy of Performing Arts, but there seems to be a particularly busy level of dance activity at the grassroots level. There are performances and workshops in studios and the streets, quite apart from conventional venues. But then Perth dance is best remembered by some visitors for the magic of performances in the Quarry and the Sunken Garden, locations that are unequalled by other cities. Buzz Dance Theatre in Perth celebrated its 25th anniversary in 2010 with its characteristic fusion of hip-hop, jazz, and contemporary dance for young people, basing its works on themes such as the environment and changing urban landscape. In 2014, it teamed up with STEPS Youth Dance Company to become Co:3 (pronounced koh-three) to extend its reach. Then there is LINK Dance Company, for WAAPA graduate dancers, giving them an experience of a professional career. And there are other organizations that provide the infrastructure for dance to happen in that city.

Dance from many cultures has a place in Australia, where there have long been traditions of local groups of flamenco and Indian classical dancers – think Anandavalli, Padma Menon, and Chandrabhanu – with increasing numbers of performers in Chinese, Japanese, and Malaysian styles. Just as there are pure forms of these dances, there are also fusions and fascinating collaborations.

And, of course, there are independent artists and freelance creators by the dozen – people who enrich the dance scene in many distinctive ways. Some are, or have been, artistic directors of companies: Meryl Tankard, Chrissie Parrott, Graeme Murphy, Sue Peacock, Russell Dumas, Nanette Hassall, Martin del Amo, and Yumi Umiumare.

The second decade of the 21st century has seen a flourishing growth of older dancers. Eileen Kramer takes the crown as the oldest performer. She turned 104 in November 2019, the same year she was making a film. She is based in Sydney and at last sighting on stage, she was accompanied by some relatively young dancers – only in their 60s and 70s. A couple of them belonged to the Australian Dance Artists, a bijou group that in recent years has made and performed work in the Sydney studio of sculptor and installation artist Ken Unsworth (born 1931). He has been the host and an integral part of the annual presentations: commissioning the music, setting the scene, and joining in the action from time to time. The dancers are Anca Frankenhaeuser and Patrick Harding-Irmer (formerly London Contemporary Dance Theatre) and Susan Barling and Ross Philip (formerly Sydney Dance Company), who were brought together by dance educator Norman Hall. They began their association in the 1990s in a series of productions that turned into larger events, memorable for their locations – notably Cockatoo Island and the Art Gallery of New South Wales – as well as their ambitious content. In Tasmania, MADE was established in 2005, its title standing for Mature Age Dance Experience. The group's prime aim is involvement and enjoyment, but this extends to having works made for them by choreographers including Graeme Murphy and founder Glen Murray, giving public performances and touring venues and festivals in Australia and Japan. Canberra has the indefatigable Elizabeth Cameron Dalman, whose Mirramu Creative Arts Centre and Mirramu Dance Company have been based at Lake George, near the nation's capital, since 1989. In 2015, she celebrated the 50th anniversary of founding the Australian Dance Theatre in Adelaide, and she continues to perform and direct. Between 2016 and 2019, she toured internationally in the cast of Michael Keegan-Dolan's acclaimed Loch nahEala/ Swan Lake. Canberra also has a senior dance ensemble, the GOLDS, which has featured in that city's BOLD festival in choreography by Liz Lea, who founded the festival in 2017.

Today's newcomer groups offer a burst of energy and fresh viewpoints. Alas, they often disappear almost as quickly as they arrive on the scene. But concerted efforts to engage audiences with dance in particular communities can have a longer life – for instance, Western Sydney Dance Action, based at the Riverside Theatres in Parramatta, with its FORM projects giving dancemakers of all kinds an opportunity.

The list could go on and on . . . and we haven't even considered the terrific commercial dancers that Australia keeps producing, whom we get to see in the profusion of musicals that entrepreneurs like to offer us. But there is one group above all others – Dein Perry's *Tap Dogs* is Australia's greatest dance export. Launched at the 1995 Sydney Festival, by its 16th year, it had notched more than 250 cities on six continents with as many as four companies touring at once. You might have thought its time had come, but in 2019, it was brought back to the stage. Maybe it isn't only Indigenous dance that can be called truly Australian.

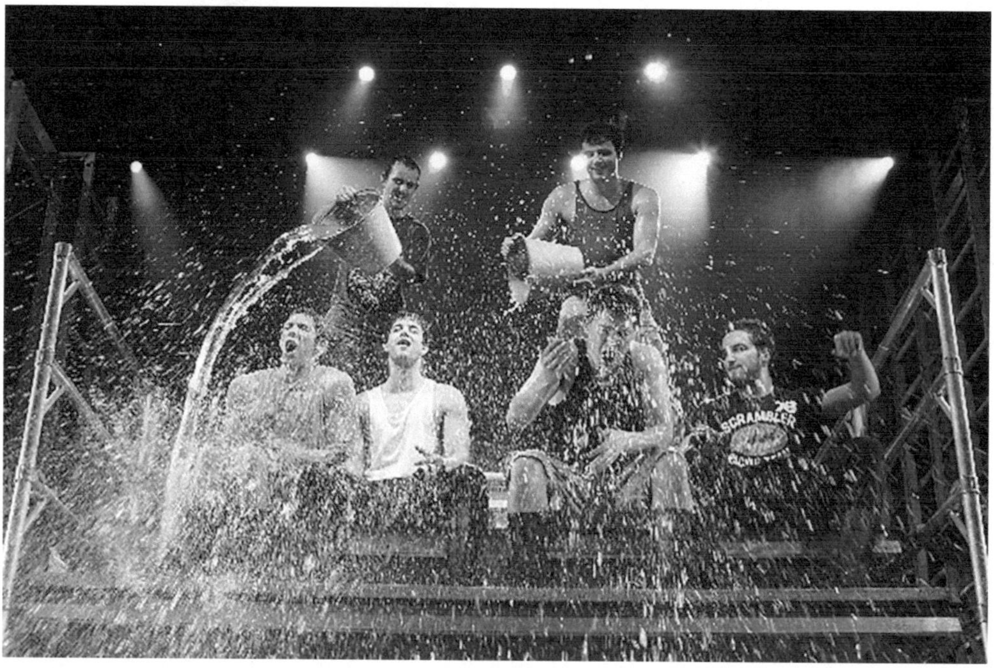

FIGURE 1.2 *Tap Dogs* (1995)

Source: Choreographed by Dein Perry

Photograph: Ralk Brinkhoff

Source: Supplied by Tap Dogs, in the collection of Ausdance National

2

TREADING THE PATHWAYS

Independent Indigenous dance

Marilyn Miller

I grew up in Cairns during the 1960s and '70s, and my sister and I were the only two Aboriginal kids in the dance school, which was run from the Harbourside Hall. I had wanted to be a dancer from a very early age, since seeing Rudolph Nureyev and Margot Fonteyn on television. I knew that's what I wanted to do. I had a dream to be a part of the first Aboriginal Islander dance company in Australia . . . that was realized many years later when I became co-founder of AIDT – the Company in 1988.

The impact of seeing dance on television is what triggered my pursuit of dance, and nowadays it is a goal I seek for other dancers and choreographers – to have their works screened on national television as well as being seen on international programmes.

AIDT – the Company was founded in 1988 by a small group of dancers who had graduated from AIDT – the College (NAISDA College) and were considering options post-graduation, with the most natural progression being to create a professional company to further our dancing and performing careers. The group included Brian (Breee-an) Munns, Lewis Lampton, Sidney Saltner, Matthew Doyle, Dennis (Dujon) Newie, and myself (as the only female), with our artistic director, Raymond Blanco.

For the first three years of its life, AIDT – the Company forged its reputation in the international arena and toured the length and breadth of Australia. During this period, the performance group Bangarra was formed by Carole Johnson, and it was also able to offer AIDT College graduates employment opportunities. For a short period in Australian dance history, there existed two dance companies providing employment for trained Indigenous dancers. At the time, AIDT – the Company was founded and was wholly run by Indigenous artists, and Bangarra was run by non-Indigenous artists, a situation that was to change later.

Those of us in AIDT – the Company dreamt of seeing dance schools and small companies across Australia that could feed into our larger national company – a satellite model – but with the demise of AIDT – the Company in 1996, the dream faded.

While the only remaining Indigenous dance company, Bangarra, continued to grow from strength to strength, considerable gaps began to emerge in employment opportunities and career development for independent Indigenous dance artists.

DOI: 10.4324/9781003160007-4

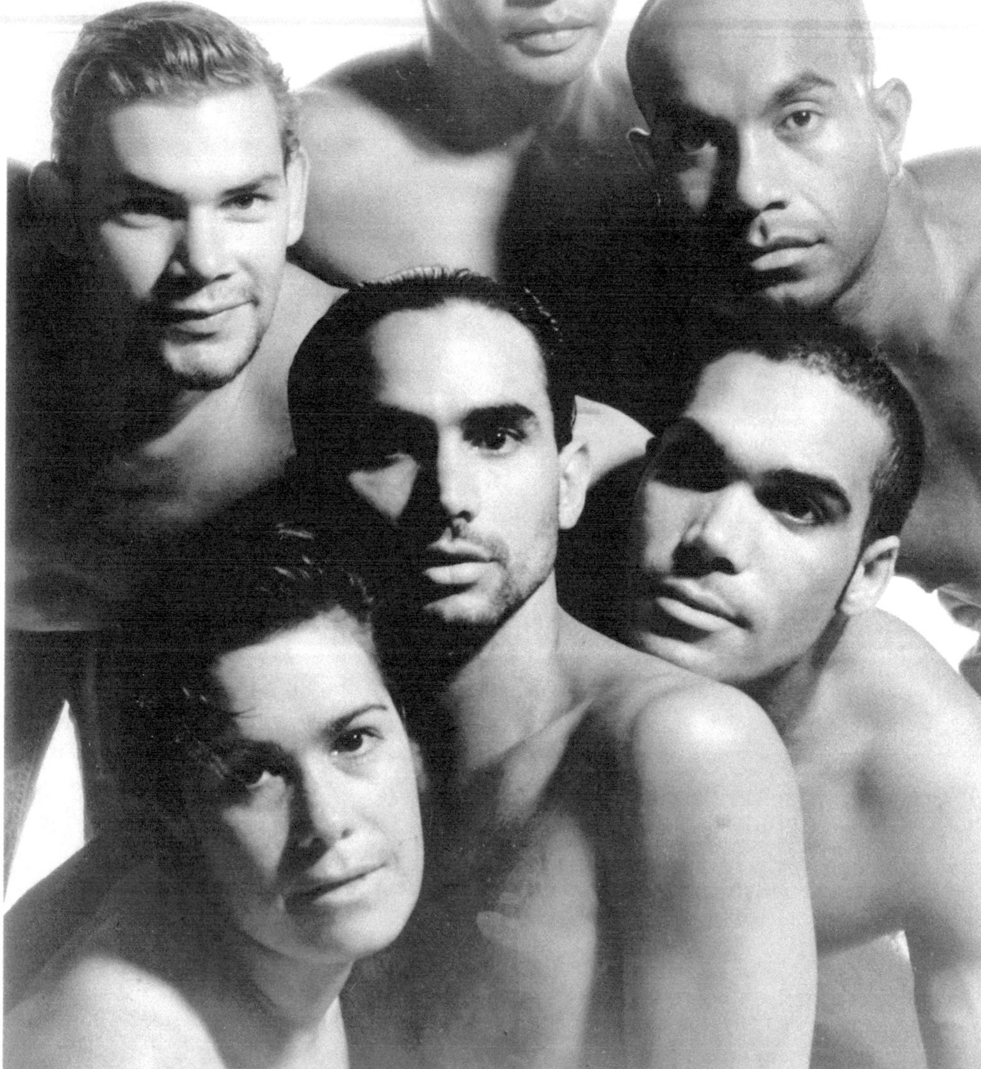

FIGURE 2.1 AIDT – the Company (1992)

Source: Raymond Blanco (artistic director); Marilyn Miller (front); Breee-an Munns, Dujon Niue, Gary Lang, Matthew Doyle

Photograph: Sally Kater

(Collection of Ausdance National)

In those years after the demise of AIDT – the Company, a significant pool of independent dancers and choreographers had been generated. With only one contemporary Indigenous dance company in Australia, the number of NAISDA graduates exceeded the limited places offered by Bangarra, so independent Indigenous dance artists such as Gary Lang, Raymond Blanco, Jason Pitt, Vicki van Hout, Samantha Chalmers, and myself were all working outside company structures.

In 2005, the Australia Council and Ausdance National facilitated the Creating Pathways national forum in Canberra, hosting over 40 Indigenous dancers and choreographers. The desire to see that dream of the satellite model realized was rekindled, and the energy generated over the three days of workshops and presentations was monumental. I realized it was the dawning of a new phase for Indigenous dance. We were now witness to a wide spectrum of practitioners, as opposed to a group of pioneers. Nonetheless, what we were about to be involved in was pioneering. As a recommendation of the Creating Pathways forum, a new position was funded by the Australia Council's Aboriginal and Torres Strait Islander Arts Board (ATSIAB), and the strategic initiative Treading the Pathways was developed in partnership with Ausdance National. I was employed as the inaugural National Indigenous Dance Coordinator in February 2007.

Initially a 12-month pilot project, Treading the Pathways was to address the issues and concerns that arose from the Creating Pathways forum. These included better networking, information sharing, employment opportunities, and the promotion and profiling of Indigenous dancers and choreographers. These were issues that needed immediate attention – independent dancers and choreographers were disheartened that their works were not being acknowledged or promoted by the dance sector and funding agencies; that their works were sometimes criticized by non-Indigenous critics and practitioners for not being Indigenous; that they were different to what Bangarra was doing. This pointed to a lack of understanding amongst the wider dance industry as to the diversity of Indigenous dance.

To bring about change required a huge shift in the mindset of Indigenous practitioners. This has not been an easy shift and is ongoing, with some reluctant to embrace and contribute to it. Others who are supported by Treading the Pathways and are being advocated for are benefiting from having their practice promoted through discussion, advice, and information sharing.

Politically, Treading the Pathways achieved a national and international profile for Indigenous dance when the national Indigenous dance coordinator was invited by the Federal Parliament to produce and direct the 45-minute 'Welcome to Country' at the opening of the 42nd Parliament in Canberra in February 2008. In a short space of time, we were able to bring together dancers from all over the country to produce a significant event in the lives of Indigenous Australians, performed at the centre of government the day before the Apology to the Stolen Generations by the prime minister, Kevin Rudd.

In the meantime, artists from across the nation had been targeted by Treading the Pathways to spearhead the strategic development of Indigenous contemporary dance in Australia; their practice is a true reflection of the diversity of Indigenous Australia. These artists, and many more, will continue to receive the support of the Treading the Pathways initiative as it transitions to becoming a new independent company, BlakDance Australia.

Gary Lang in Darwin, a Larrakia man, creates works that have strong classical influences as well as strong cultural influences. Since October 2008, when the national Indigenous dance coordinator approached the Darwin Festival, Gary has enjoyed the support and producer services of the Darwin Festival and has presented his works over the past three years in tandem with the Festival and the Darwin Entertainment Centre. In August 2010, *Goose Lagoon* was premiered at the Festival to full houses and positive reviews.

Gail Mabo is a descendant of the Mer People, Murray Island. In 2005, Gail presented a work-in-progress showing of her choreographic debut work *Koiki*, based on the memories and stories of her late father, Eddie Koiki Mabo. Gail then presented the 20-minute work at the Dreaming Festival in 2007 and 2008. In 2009, a review of the work was undertaken, and the report produced several suggestions and recommendations for the further development of the project. One of these recommendations was for a producer to oversee and assist the administrative needs, and another was for a dramaturg to assist with the storyline – the national Indigenous dance coordinator is facilitating this networking and relationship building.

Rita Pryce is of the Kulkalgal people of Gaigalkuth on Poruma (Coconut) Island and resides in Cairns, Far North Queensland. Her first full-length work, *Warupaw Uu*, was presented at the Dreaming Festival 2010 and then an excerpt presented for the Cairns Indigenous Art Fair 2010. She was also co-choreographer for Gail Mabo's *Koiki*.

Vicki van Hout is one of the most prolific creators of Indigenous contemporary dance, presenting work that challenges stereotypes, questions preconceptions, and incorporates iconic Indigenous and non-Indigenous motifs. Vicki has worked with youth theatres, professionals, and theatre practitioners, her theatre background allowing an easy transition when working between art forms. Since 2009, the national Indigenous dance coordinator has been sourcing producers for Vicki's work.

Nicki Ashby is also included in the targeted artists programme of Treading the Pathways, with a placement with a leading training institution currently being negotiated. This placement, like the company placements that Gary Lang, Rita Pryce, and Vicki van Hout have enjoyed with The Australian Ballet, Expressions Dance Theatre, and Chunky Move, respectively, will allow Nicki to gain insight into organizational structure while drawing comparisons with career goals.

Indigenous practitioners who have contributed to the growing independent sector and who have also been involved in activities that Treading the Pathways has assisted, supported, and/or initiated include Deon Hastie, Nikki Ashby, Albert David, Djakapurra Munyarryun, Matthew Doyle, Rosealee Pearson, Henrietta Baird, Genoa Gela, Peggy Misi, Shellie Bin-Garape, Darren Edwards, Tara Robertson, Ses Bero, and George Dowd.

Along with choreographers who have enjoyed the support of Treading the Pathways to develop their practice, graduates of training institutions are now presented with a selection of employment opportunities with Indigenous creators around the country. This is affording them prospects that more established choreographers were unable to experience in the past.

We are now seeing the formation of a network of dance groups and companies that those of us in AIDT – the Company dreamed of – the growth of the satellite model. This network now needs collaboration between these groups to showcase the diversity of Indigenous dance and highlight regional differences in dance from around the country.

And, if another dream is to come true, then perhaps that showcase will be televised both nationally and internationally.

Note: This chapter was originally published in Burridge, Stephanie and Dyson, Julie (ed.) (2011) *Shaping the Landscape: Celebrating Dance in Australia*, Routledge, India.

Cambodia

3

I AM A CAMBODIAN CLASSICAL DANCER

Hun Pen

To my parents, Hun Sarin and Heng Nayto, and to my teacher, Chea Samy

It is a widespread belief among Cambodian people that Cambodian classical dance is a simple and pleasurable skill and has little value in relation to the other skills in society. Yet classical dance is an art form that is becoming increasingly prominent in Cambodia. It reflects the Khmer identity and has gained attention on the international stage as a unique art form due to its technique, integrity, and elaborate costumes. However, outsiders may not realize how important it is to the people of Cambodia – classical dance is one of the art forms that depict the verve and the colour of the Cambodian people. To ordinary people, dancers and their dance technique may look the same when they perform on the stage. However, it is important to describe the training, and how one trains, in order to become a good dancer. There are many facets of my life that are intertwined with classical dance and art: travelling, education, academic studies, political involvement, dance training, leadership training, and drama, as well as voice training. I will now give a brief description of a small dimension of my life in dance that naturally included high and low moments. This chapter covers the highlights of my life as a Cambodian dancer. I will divide this autobiographical sketch into four brief sections. The first is how I started to study dance, the second is my training in classical dance, the third is my transitional period, and the fourth is the conclusion.

I am a Cambodian classical dancer. I am the youngest daughter in a family of four daughters. My father is a leader and a man of compassion, tolerance, patience, and understanding, and my mother is a beautiful, loving, caring, dedicated, gentle, respectful, and strong woman. I grew up in a strong family where my father has always been my intellectual advisor and teacher, while my mother has always been my mentor and guide. Their dedication, devotion, patience, and guidance have played a major part in helping me to become who I am today – a dancer and educator. My life as a dancer on the stage may appear to the audience as a prima ballerina who is happy; my friends and colleagues see me as a woman with luck. Yet they may not know how I have struggled with my dance training and my education – to be a good dancer in a very competitive field, I have had to face many struggles and challenges, both emotionally and physically.

DOI: 10.4324/9781003160007-6

Five years old and my first stage performance

I met my dance teacher who later (in 1990) became my personal dance teacher when I enrolled at the School of Fine Arts. Her name is Chea Samy. She was a female role dance star and was the best female dancer in Cambodia – known in Cambodia as a 'living national treasure'.

I recall the time when I was five years old. I often followed my mother to the dance hall where she taught folklore dance, and along the way, I saw so many dancers rehearsing. Their dance movements were so beautiful and elegant. I watched the dance training at the School of Fine Arts almost everyday. I loved to see them dancing, and I loved to imitate the dance when I came back home. I met Master Chea Samy at the dance hall where she taught the students. She saw me standing with many other kids by the window watching the dance training – she called me in and asked me with a big smile if I wanted to learn how to dance. I recognized her as the most famous dance teacher, and so I was happy to say 'yes'. She taught me a few dance movements. Very quickly, she said to me, 'You are very good and you should learn how to dance. Come back again tomorrow and I will teach you more . . . tell your mother'. I did not respond, but I knew that I wanted to learn more about dance.

My mother came and picked me up that day. I heard Master Chea Samy tell my mother how good I was and that I should study dance – there was no response from my mother except her smile of respect to Master Chea Samy. However, once I got home, I was upset. I heard my mother tell my father about Master Chea Samy's comments. My father obviously did not want me to become a dancer, and neither did my mother. I did not realize how much I wanted to be a dancer until the few times I returned to the dance hall. I received a great deal of attention from Master Chea Samy. One day, Master Chea Samy taught me a dance that was an episode from *Preah Chinavong*. I kept coming back for about a week and practiced the dance with her. Then she asked me to dance with the music. A group of Pin Peat musicians played the music, the singers started to sing the song, and so I started to dance following the music and the song. All the people in the dance hall stopped and watched me. I saw Master Chea Samy had a huge smile on her face. I felt so proud, but I was also shy, and when I finished the dance, she told me to come back to rehearse the dance a few more times before performing it on the stage.

I was only five years old, but I still remember how it felt to hear the words, 'perform on the stage'. I was so happy but at the same time nervous. A few days afterwards, I performed on the stage as a young Preah Chinavong with a very nice dance costume. I was so proud and got admiring comments from dance teachers as well as audience members. From that moment, I realized that I really wanted to be a classical dancer. Even though I knew my parents would never want me to become a dancer, I felt strongly that I would do anything to become a dancer. I told my parents that I wanted to study dance. They told me that they loved me so much and that they cared for my future, and they did not want me to face difficulties when I grew up. Not understanding their decision, I remained firm in my mind, although I continued my academic studies at the public school.

A few years later in 1990, at the end of fourth grade in school, I heard about dance auditions at the University of Fine Arts. I escaped from my school and went to watch the dance training. I escaped not once but a few times until the teacher reported my absences from class to my parents. Knowing that I was always the best one in the whole school, my parents excused me for missing classes. Soon afterwards, I enrolled for the dance audition at the dance school. It was three days of dance auditions, dance examinations, and general knowledge. I passed the dance audition and examination with an excellent grade. I enrolled at the University of Fine Arts as a student from that time, although my parents were still not happy about it.

Private lessons with Master Chea Samy

I came to study at the School of Fine Arts six days a week, where I did dance training in the morning from 7:00 until 11:00 a.m. and studied general knowledge from 1:30 until 5:00 p.m. according to the school curriculum. Four hours of the dance training was routine; every morning we would assemble and sit in rows according to the dance role we were expected to perform. I was selected to be a 'female role' in classical dance and sat in the first row on the right-hand side, which was the position for the rhythm leader of the group. Reciting the 4,500 movements, the extensive vocabulary of Cambodian classical dance, was required at the beginning of every morning of practice. From start to finish, this exercise would take approximately one and a half hours. Thereafter, we would be separated to receive specialized training according to our dance role.

After one month of dance training, I was the only student of my generation to be selected to perform with the older generation of dance students, which made me feel proud and surprised. Soon after, I was selected to be Master Chea Samy's top student, so I was very honoured. There were only four of us out of the hundreds of students at the University of Fine Arts who were selected. My heart felt so proud to know I had a great opportunity to have private lessons at her house where I could learn extraordinary and special dance secrets. These techniques were not being taught at the school, and many students did not get a chance to learn them. These private lessons at her house were conducted in the evening for two or three hours after I finished my general knowledge at school.

These extracurricular lessons by 73-year-old Master Chea Samy signified her devotion to the preservation of classical dance – they went beyond the rudiments of dance movements and delved into the intricacies of classical dance composition. As the most senior and accomplished dance master, Samy was able to transmit priceless pearls of wisdom and training advice to me. Every evening after school, I would go to her house and meet in her living room dressed in *samput chanken* and *aue lakhaon* (a short-sleeved shirt for dance training) while Master Chea Samy sat in her armchair and conveyed the movements to me.

Despite her ill health at the time, she committed herself to teaching her young charges the secrets and subtleties of Cambodian classical dance. Only when necessary would she actually rise from her chair to demonstrate a particular technique. I was made to repeat each movement in detail, commit it to memory, and perfect it, after which Chea Samy would pay particular attention to my facial expressions, body, legs, and hand movements. She demanded precision at all times. At the end of the technical component of each practice session, Master Chea Samy would explain the dance techniques in the context of the history of classical dance. This was an attempt to situate the importance of classical dance within the socio-political movements in Cambodia during her time and to pass on this understanding to me. Her perseverance and hard work paid off, as I was able to give a successful performance to the public at a very young age. This was often a painstaking and time-consuming process, but it was an invaluable transmission of knowledge that was lost when Chea Samy passed away.

The last memory I have of her still stays in my head, and I still feel badly about it today. After a photo session one day, Master Samy called me to her house. When I arrived in the evening, I saw her sitting in her own armchair where she usually taught me. The room was a little dark, so I could not see her face clearly. The only things I remember so deeply were her words to me with a sad face. 'Come in, I do not know how much longer I can live to see this world. You are my hope and I trust you a lot that you will continue my work – to carry this heritage and spread it to the world and to the next generation of students', she said to me. I heard what she said, but I was too young to take it seriously. The last dance lesson she taught me was an extraordinary dance known only by two dance masters, Master Chea Samy and another master – I am still searching for her. She taught me despite her pain, which I did not understand. She wanted me to remember the song, as it was important to preserve it as much as the dance technique itself. She made me dance and sing the song at the same time. Even though I was out of breath singing while I was dancing, I still had to do it. The lesson went beyond the sadness and emptiness that I felt inside.

This situation, with her devotion to my training, stopped two weeks later due to her emergency surgery. Unfortunately, I could not even talk to her before she passed away a week after the surgery. As my mother wanted to continue to support my talents and skills, she became my dance teacher after Chea Samy passed away. She gave me private dance lessons at home. She taught me special techniques for classical dance that I was not able to learn from school, for instance, how to use my emotion and spirit with my body – that was so hard to understand and practice. She had me dance while she observed the dance movements and explained to me how to use my spirit and emotions along with my hand gestures. She taught me how to use my eyes with my spirit and the breath of my hands while dancing. It was the most extraordinary technique I have ever learned and experienced.

Performance experiences

From my training with Master Chea Samy and her direction, I was able to give a variety of remarkable performances at a very young age, both nationally and internationally. Besides a busy schedule at school and dance training, I often made time at night to perform for public and cultural events as well as government functions and at embassies. My role as 'Sovanmacha' (golden mermaid), an episode from the *Reamker* story, was well known. My first performance of this role was at the Russian embassy when I was eight years old. I was dressed in my costume for almost 45 minutes and was not allowed to use the bathroom after the costume was in place. Thereafter, I would put makeup on in order to paint my face a certain colour and shape. This would take around another 30 minutes, after which I would be free for a few hours until the preparation session before the performance started.

Between the free hours of my preparation for the performance, I often warmed up, rehearsed the dance, and spent time on my academic assignments for school. Around one hour before the performance started, I would prepare my hair so it would be ready to place the head crown on it that weighed around 1 kilo. According to tradition, after being crowned, I would pay respect to my teacher and the ancestral spirits by offering five incense sticks with two small candles to my dance master. This was done to seek her blessing for a successful performance. Subsequently, I would train myself to concentrate and meditate for 15 minutes before stepping out onto the stage. I have followed this pre-performance routine for the entire 20 years I have been performing in Cambodia and abroad.

After one year of training and experience in Cambodia as a leading dancer, I was selected for my first performance tour to France. I was nine years old at the time and too young to have any idea what France would be like. I thought it would be a country full of white people with long noses. Our audiences were packed for the whole month and a half each time we performed in different cities. To relax after the performances, I engaged in activities with the other dancers such as workshops, sport, and language exchange. I had a chance to visit cultural sites, museums, royal palaces, and many other interesting places. As a young dancer, I merely enjoyed and absorbed the experiences. A few months upon my return to school, I received the award for the most outstanding student and dancer from the school. For that reason, I had to study even harder and get more involved with the newspaper, photo shoots, and other film projects. From that point onwards, I had to work hard to manage my time between my studies at public and private school as well as my performances and other cultural events. Sometimes, I had to run very quickly from a seven-hour photo shoot or performance with bare feet under the hot 39- or 40-degree Celsius sun to my classes and submit myself to a two-hour academic exam. Then I continued on to another academic private school for two hours before going home to have a late dinner. This was done with my dedication, patience, and devotion as well as with great support from my parents.

I kept growing up, and when I reached puberty, my body transformed and changed. Hard working and accomplished, I was acknowledged by the school as the most outstanding student and dancer every year. Later, in 1997, I was rewarded by the government as the best student and dancer. As a well-known dancer, I was invited to study Indonesian dance to perform for the Indonesian leader Suharto and the king of Cambodia at the royal palace. Seeing my talents, the princess of Cambodia requested me to perform solo in Chinese dance, Vietnamese dance, and Indian dance. By this time, my name in the world of Cambodian performing arts had become extremely well known. Regardless of my popularity, I continued to enjoy participating in classical dance tours, conferences, and workshops in various countries in Europe and Asia. However, in 1997, the feelings of exhaustion from all the travelling began.

My experiences of the performance tours in Europe, which I would call 'hectic', were a mixture of joy, sadness, and lack of sleep and rest for the whole two and a half months. The group and I had to work more than 17 hours a day at times. We were only able to sleep in bed for four hours or so, and the rest of the time we spent in transit on the bus or at the theatre. We had to check out of the hotel to get ready to be on the road by 7:00 or 7:30 every morning. During the 10-hour bus rides, we would try to rest and enjoy each other's company. I never could fall asleep on the bus, as the seat was too small and not comfortable. Therefore, I often read books and tried to do some assignments I brought with me from Cambodia. Normally I could not accomplish much. Around 12:00 or 1:00 p.m., we would stop at a small restaurant along the way for about one hour for lunch and then continue our journey. We would reach the theatre at around 5:30 p.m. and rush to get off the bus where we were able to bring only our makeup and would leave our suitcases on the bus. We rehearsed for about an hour and then started to put on our costumes and get ready to perform at around 8:00 p.m. The performance would finish almost two hours later, and we were able to undress and get ready to leave the theatre at around 11:00 p.m. From here, we had to spend another one or two hours on the bus to reach the restaurant where we could have dinner at around 1:00 a.m. By the time we finished dinner, it would be 1:45 a.m., and by the time we checked into the hotel and got ready for bed, it would be 2:30 a.m. or so. Sometimes we would do two performances, one in the afternoon and the other at night.

This demanding schedule for the whole two and a half months made me feel very homesick and exhausted. One good thing that helped me to relax during the tour was that I could visit and see the different countries in Europe. Despite the hardships and difficult tour schedule and tiring travel, I did not want to give up dancing. I continued to participate in other performance tours up until the point where I felt I had done enough performing and I needed to engage in other interesting activities beside classical dance, such as contemporary dance workshops or other conferences.

My transitional period

Having learned and gained so many experiences in the performing arts for 11 years and after having made my reputation as a student and a dancer, I was selected as president of the faculty of students in 2000–2001. As a young woman in the field of dance, I had to be involved with and work closely with government leaders and leaders from other universities in Cambodia. During the two months' campaign in the city, I had to participate, meet important people for discussions, and cooperate in order to promote the arts that were not so well known in Cambodia at the time. My campaign was successful, and I was able to convince the government and the intellectual community to support and promote the performing arts as well as providing opportunities for the poorer students at my school. With the funding support from the municipal city, bags of rice and bicycles were given to the poorest selected students at the Faculty of Choreographic Arts. Various performing arts activities were organized for the benefit of the public as well as for other audiences from universities throughout Cambodia.

Starting that year, I distanced myself from performing classical dance but became more engaged in arts management and involved with organizing performances in the country. Nevertheless, I travelled whenever I had the chance. In doing so, I became exposed to changes in society and to other cultural differences. Soon after that, I took an interest in contemporary dance. My grounding and experience in classical dance and my experience from learning different dance forms of other cultures has allowed me to become skilled in contemporary dance relatively quickly. I participated in various contemporary dance workshops, classes and conferences in many countries in Asia and Europe and the United States – moreover, I became a successful choreographer and performer in Cambodia and other countries in Asia. In spite of my successful dance career, however, I did not overlook my academic pursuits. Having realized the lack of written documents on Cambodian art and dance, I started to pursue my academic research in 2005 on dance and its history. After four years of academic research, I have gained a greater insight into and understanding of Cambodian culture and dance in the socio-political and historical context.

Conclusion

Many people have asked me what makes me continue dancing as a Cambodian classical dancer. Why is the soul of classical dance so important to me? Why does dance mean so much to me? It seems to be a simple question, but it is difficult to answer. The simple answer I would give is, 'because I love it and I am passionate about it'. I have been dancing for so long, and I know this dance form so well. This dance form has become rooted in my heart and mind. It is alive within my blood. I have been devoted to the art and have exercised my talents to present what I, as a Cambodian, have to offer the world. Dance is part of my identity and the talent that I have – to come this far, I have undergone so much hard work in training and studying with my dance master, Chea Samy, and my mother. It is so important to have a very good dance master in order to train and to take pains to produce one good dancer. For this, I owe tremendous gratitude to them. My foundation in classical dance has complemented my training in contemporary dance movements. Despite the fact that I am pursuing my contemporary dance career and academic research, I still identify myself as a classical dancer. To develop myself to be a true dancer and performing arts scholar is not easy. I am trying to reflect and draw the connections between what I have learned and the experiences I have gained in the three areas of classical dance, academic research, and contemporary dance.

Note: This chapter was originally published in Burridge, Stephanie and Frumberg, Fred, (ed.) (2010) *Beyond the Apsara: Celebrating Dance in Cambodia*, Routledge, India.

4

DANCE EDUCATION IN CAMBODIA

Chey Chankethya

Cambodia is one of the oldest countries in the world – it used to be a great empire, especially from the 2nd to the 15th century. The Angkorian Empire spread its power and civilization across the territory, and even from this time, dance was an important aspect of the culture and tradition of Cambodia. Dance played an important function in Cambodian society, and although the function has changed over time, Khmer dance has a unique way of transforming itself from generation to generation. This chapter will relate how dance is passed on from one place to another and from one person to the rest. Because dance has survived in different places and is performed by people with different levels of skill, transmitting it is a complex task. Combined with the lack of written documentation or dance notation, transferring through the body can result in confusion and uncertainty. However, because of its importance, Cambodian dance has survived for more than 2,000 years – to find out more about its origins and sustainability, we need to probe further.

The origins of Khmer dance

From the sculptures that were found in Takeov Province, for example, Hindu divinities such as the eight-armed Vishnu that date from the 1st–6th century,[1] we might assume that dance served mainly a religious purpose during this period. At that time, people also believed in animism and offered dance and music to appease the spirits and as a tool to solve their problems. For example, when people needed rain to do their farming, they danced to ask for it from the gods – dance and music were also used to cure people from illness. Because dances were created for spiritual reasons, the movement and gestures were not set, and the dancers just moved their bodies according to the rhythm of the music. Over time, different kinds of dances and art forms were created to serve people's needs. Dances changed according to different beliefs and when people relocated from one place to another.

The role of dance

Despite many difficulties throughout the history of Cambodia, several factors enabled dance to survive until the present day. One reason is the changing role of dance in society – dance

DOI: 10.4324/9781003160007-7

managed to adjust itself to any period because of its beauty, value, and importance. From the earliest time, dance was believed to be a conduit between the gods and human beings, or between heaven and earth. In the past the king, as well as the common people, offered the sacred Khmer dance as a gift to please the spirits or gods. For example, most of the classical dancers during that time were female and were required to keep themselves pure in body, mind, word, and attitude. They regarded themselves as the god's messengers, and because of their purity, they were able to talk to the gods – this set them apart from the ordinary people.

However, depending on the time and situation, dance served other purposes – it evolved from having a sacred role to having an entertaining role for the common people. Audiences loved to watch dance because of its refined, smooth movement and the beautiful gestures that were accompanied by songs and instrumental music. With every movement, the dancers made their audiences feel as though they were in heaven, and it left them with a sense of peace and serenity. Beyond the beautiful and perfect movements of the dancer, the spirit

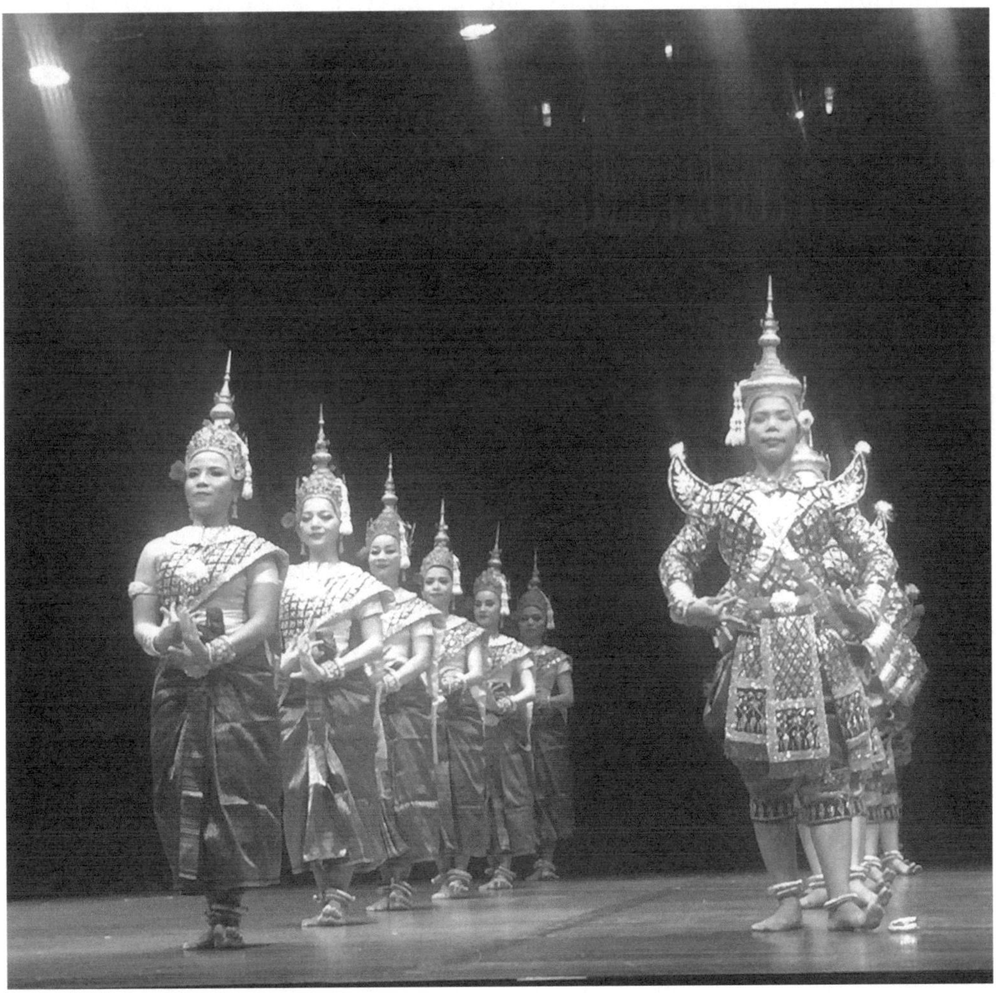

FIGURE 4.1 Performance: Robam Preah Thong

Source: Photograph: Author's personal collection

of the Khmer people was revealed – thus, at that time, dance was regarded as a valuable and important attribute of national identity.

Much later, when dance was brought into the schools, it assumed yet another role. People learned dance at school as a subject and were able to pursue a career in dance and earn a living through it as a profession. This current role is an important development in the evolution of dance in Cambodia and a reason it is sustainable today.

Learning dance

Traditionally, Cambodian dances were passed on from one person to the next – there were no formal classes or schools for the teaching and learning of dance. Moreover, it may not have had a clear structure of movement or gestures, because in the beginning, people danced within a ceremonial context. They chose simple movements that enabled the dancers to move easily to the rhythm of the music. Therefore, the teaching and learning process happened in a very simple way – probably through family groups transferring the dance to each other. However, following the influences from Indian court dance (possibly as early as the 8th century), a clear structure of movement was established in Cambodian dance.

From the 5th to 19th century, dance was very popular amongst the royal families and dignitaries. During this period, dance had an intimate association with the royal palace because it was the place where religion was highly respected. Dance was not seen by the common people – only by the king, who was considered a god, and some other important people who worked for him. Dance and the dancers were regarded as the property of the king, which meant that dance could only be taught and learned within the palace compounds. In addition, temples were also places where dance could be performed for religious purposes.

Not everyone was able to watch or to participate in dance. Mostly dancers were members of the royal family, the king's servants, or girls from elite families. It seemed impossible for dance to extend outside of the great wall of the royal palace. However, this closed gate was opened in 1431 A.D. when the Angkorian Empire collapsed under siege from Thai invaders. Dance was then brought to different parts of the country by the palace dancers. Some dancers stayed with the royal family, while others spent their lives with different government officials, the pagoda, or a community. Dance during this period not only belonged exclusively to the king but also to anyone who wished to assemble their own troupe of dancers. However, in the 19th century, during the reign of King Ang Doung, once again there was a dance troupe at the royal palace. The dancers lived a secluded life inside the palace walls, practicing daily and performing for the king. They normally performed for religious purposes or for royal occasions such as coronations, weddings, and important events. Dancers normally were the wives of the king, his relatives, or servants – no one outside of the palace was allowed to learn dance and to perform with this troupe.

However, there were also instances of court dignitaries who supported dance troupes. For example, in the early 1800s, the governor of Battambang province reportedly had both male and female dance troupes to serve his family's needs. He usually entertained himself by watching dance or listening to the music of his own troupe. On special occasions, he also allowed the people in the province to enjoy the performances; however, similar to the royal palace troupe, the governor's dance troupe did not allow outsiders to learn dance or perform in this group – only his relatives and servants could learn dance and perform.

Apart from the royal palace and family compounds, pagodas were also places where dance could be taught. For Cambodians, pagodas have been places where religion, culture,

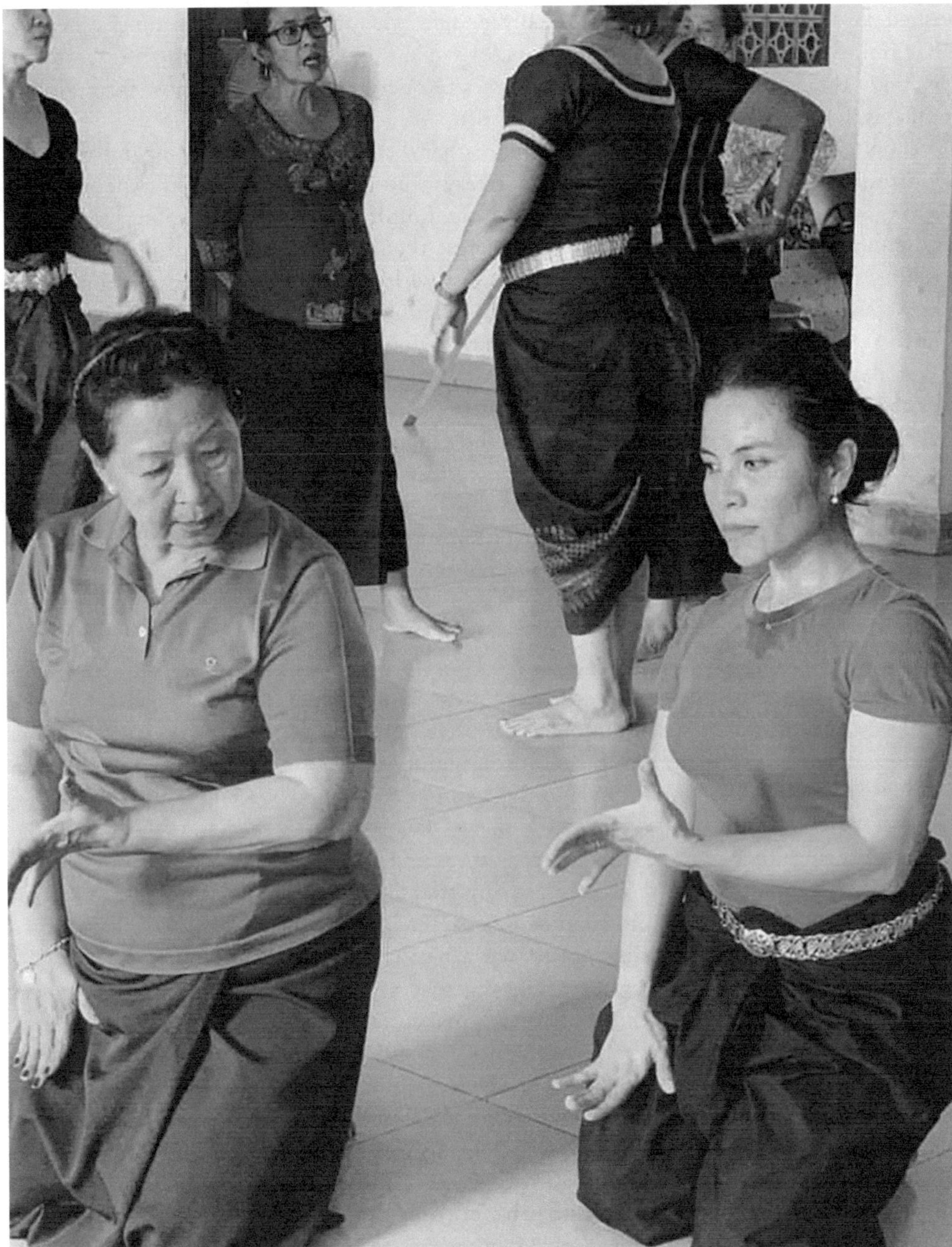

FIGURE 4.2 Chey Chankethya training with Master Menh Kossany

Source: Photograph: Author's personal collection

and tradition are strongly embraced by the people. Unlike the royal palace, the pagoda provides an equal opportunity for everyone to learn and practice their religious beliefs as well as learning about their culture – pagodas had their own dance troupes. Through the support of the monks, informal dance classes were conducted at the pagoda for those who wished to dance for the troupe, especially people who lived in the community nearby. They trained only when they had free time or when the pagoda needed them to perform. Currently there are few of these pagoda classes in operation, as they are now facing many problems; for instance, lack of financial support.

Although audiences could see dance more than before, there were comparatively few opportunities for anyone to learn to dance. Everywhere seemed to be closed to people who did not belong to an important family or a community – despite this, some indigenous communities created and taught dance amongst themselves. Thus, we can see that the teaching and transforming process happened just within a particular group of people who had a close relationship with each other. For example, from parents to children, from a boss to a servant, or across people who shared the same compound – but less from a teacher to a student. As a result, although dance occurred throughout the country, there was neither accurate teaching nor documentation of the dance. Traditionally, Cambodian people preferred to teach one another by telling or showing the gestures and movements without consulting written documents. However, in retrospect, if there had been written documents, they would probably have vanished during the war, been stolen, or been hidden away somewhere. Through the tradition of transmitting it orally and physically from generation to generation, dance almost disappeared on a few occasions throughout our history

Dance institutions

As previously stated in this chapter, schools for dance and the arts have been in Cambodia for a long time. However, the common people were not aware of this type of dance training because it occurred in the royal palace or in community settings without a formal curriculum or standards that people could learn – so this was not considered a 'school'. For example, in the past, the kings always had their own troupe of dancers who learned and rehearsed in the palace – but since they did not use a curriculum or understand that they were training to be dancers, they did not realize they were actually attending a 'school'. According to Georges Maspero,[2] who published a photograph of a dance rehearsal at the palace Chanchhaya dance hall showing a dozen dancers being trained, one can conclude that the palace school was established a long time ago. However, it was not regarded as a school because not everyone was allowed to attend the classes, it did not have a formal curriculum, and it belonged to the king.

During the period of French colonization (1863–1953), almost everything changed. In 1919, the idea of having a fine arts school had begun to emerge. There were many reasons for these changes, and they affected all aspects of life in Cambodia – central to this was the introduction of modern administrative reforms. During this era, the power and wealth of the king diminished, and this caused things to progress in a different way; for instance, the palace dance troupe became weak and unsustainable at the royal palace. In 1927, after completing his coronation ceremony, King Sisowath Monivong signed an agreement with the French to transfer the royal dancers to the authority of L'ecole des Beaux Arts. He accepted this option as the only way of saving the troupe – he also noted that the current method of teaching by transference could jeopardize the future of the dance, so the notion of a school evolved.

L'ecole des Beaux-Arts was founded in 1919 with the mission of reinvigorating some of Cambodia's fine arts and crafts. It was also a precursor to the conservatory of performing arts. In 1964, the Universite Royale des Beaux-Arts was opened. It had a mission to foster indigenous scholarship of the arts. Many different kinds of dance and art forms were brought from the provinces to be studied at the school – to be re-choreographed or re-arranged without losing the original attributes before eventually being performed for the public. Furthermore, dance was considered a subject in the curriculum. The school was regarded as the central place where dance could be preserved, and through the school's efforts, dance was promoted to national and international communities. A formal school curriculum was set, classes were conducted regularly, and the school gate seemed to be open for everyone in the country from diverse levels of society. In the morning, students were required to attend dance technique classes, while in the afternoon, they had to study general knowledge. The teachers were locally educated people, with some graduates from abroad, especially France. However, for classical dance, the masters continued to be supported by the royal palace.

During this new reformation period, the school had a strong alliance with the government, the French authorities, and the royal palace. Each of them supported the school in different ways. For instance, despite the school being open, the classical and mask dance forms were learned at the palace under the supervision of Queen Kossomak, while folk dance classes were conducted at the school outside the palace. Remarkably, the Queen was the one who was aware of the dangers of not being able to sustain dance, especially the classical dance form. She was an important person in Khmer dance history, and she devoted herself to preserving and developing classical dance after King Aug Duong's reign. By combining with the other dancers from King Sisowath's reign, the Queen was able to form a troupe of classical dancers and later provide classical classes in the palace compound. In 1965, there were 30 classical dance masters and 500 students recruited from all social classes. Since it was established, the Université Royale des Beaux-Arts was very successful in producing high-quality artists to work in the arts industry in Cambodia. Besides this, the school also played an important role in compiling documents and books about the art forms, and these were available to the students for their research. With the supervision of the Queen and the quality of the school itself, dance managed to re-emerge, and its popularity extended over a huge territory.

However, something seemed to change after the coup in 1970 led by Lon Nol.[3] The school once again changed its name – this time to the University of Fine Arts. At first, everything worked as usual from the beginning of the new government, but later, all the school palace teachers and students were moved to the school, and classes were no longer taught at the palace. In 1975, the school was closed and abandoned completely when the Pol Pot regime arrived in Phnom Penh. Immediately the country was torn apart by the war. There were no more dance masters or dancers in the dance hall at the royal palace. Similar to other schools in Cambodia, the university remained quiet, and training was abandoned. Dancers had no time to practice their dance sequences or to perform on stage – they were required to work tirelessly in the heat doing jobs that were very different from their dance training. Living in fear and starvation, the dancers and dance masters had no time to think about dance. During the regime, there was no dance education in Cambodia. Millions of well-educated people were killed, including dancers – many also died of disease and starvation.

Royal University of Fine Arts

At the end of the Pol Pot regime, the corps de ballet was reinstated under the direction of the senior masters who had survived the Khmer Rouge period, and the school was re-opened in 1980. It was named the School of Fine Arts – this institution was regarded as a secondary school. There were five skill-based schools of the arts: plastic arts, drama, dance, circus, and the school of music. In 1988, once again, it was named the University of Fine Arts, and in 1993, following the restoration of the monarchy, 'Royal' was added, making it now the Royal University of Fine Arts (RUFA). Despite many difficulties, the university is open to all applicants who would like to aspire to a profession in the arts. Anyone with an interest can apply to the school to study.

There were five different faculties – the faculty of architecture and urbanism, archaeology, fine arts, music, and choreographic arts. However, the Faculty of Music and Choreographic Arts was closed down for a period due to the shortage of instructors, and it was re-opened in 1999. In 2000, RUFA was accredited as a public institution that allowed both scholarship and fee-paying students to attend the classes. In order to provide students with a quality learning experience, RUFA often arranges exchange programmes within local and international communities so that they can share and learn from one another. Numerous masters and students have been sent both abroad and to different provinces in the country to study and undertake research and to bring back knowledge for themselves as well as for the benefit of RUFA.

As one of the most important educational institutions in the Kingdom, RUFA has been supported by different partners in order to improve its educational standards as well as to preserve Khmer traditional culture. Under the administration of the Ministry of Culture and Fine Arts, RUFA has an obligation to develop professional dancers, arts managers, resource personnel, and art advocates who will be able to work at the ministry as civil servants after finishing their study. Once again, RUFA has a strong relationship with the Royal Palace, where the emphasis is on preserving and maintaining traditional culture. There are many members of the royal family who have devoted themselves to the preservation of the traditional Khmer arts, such as King Aung Duong, the King's mother the late Sisowath Neary Roth, Princess Norodom Buppha Devi, and many more.

Although the school was already open with a clear curriculum in place, RUFA has a responsibility to work carefully on behalf of the entire nation to preserve and revitalize its culture. During the '90s and 2000s, research and documentation of the old materials that were facing extinction became possible with sponsorship from the Rockefeller-funded Mentorship programme through the Asian Cultural Council. Through this program, RUFA was not only able to rearrange the old dances that were kept alive in the old masters' memories but also to document many of the art forms, especially classical dance. To the present day, the mission has not been completed. Under the supervision of the Ministry of Culture and Fine Arts, both dance masters and younger generations are working hard together documenting the dance form for future teaching and learning resources. They are documenting the past for the future.

Faculty of Choreographic Arts

At present, RUFA is facing new challenges. In accordance with the mission of preservation, while keeping pace with current educational expectations and standards, the Faculty of Choreographic Arts was re-opened in 1999. The faculty focuses on giving students a new kind of knowledge that is different from what they have learned in the school of dance, where the focus is on following the masters' footsteps and studying the old traditional masterpieces. In contrast, the students learn to choreograph new pieces of dance, or any of the other art forms, through their basic skills in dance, theatre, or circus. Throughout the learning process, students can generate their own ideas and explore their personal creativity.

The Faculty of Choreographic Arts aims to teach the students the notion of creating and arranging their own works while having a strong sense of responsibility and a professional work ethic. Although the term 'choreographic' is a new concept for the traditional artists, it has created a lot of interest amongst the younger generation at RUFA.

Secondary School of Fine Arts

This school used to be under the same roof as RUFA; however in 2006, the school was renamed the Secondary of Fine Arts, and it was officially certified independent from RUFA. In the school, there are five different departments – these are dance, music, theatre, circus, and fine arts.

Initially, there were not many students, but numbers have increased over time. Children aged 6 to 13 were selected to attend to study in three different dance forms according to their appearance and ability. They have to spend nine years in order to complete their academic training, and this includes two major exams. Unlike the Faculty of Choreographic Arts, the Secondary School of Fine Arts trains students to become teachers and performers of the traditional dances. They are required to learn all the basic techniques and theory; in addition, they are only taught the dances and dance stories that were choreographed a long time ago.

Because it is the only place where the arts can be professionally taught, the Secondary School of Fine Arts plays an important role in maintaining the traditional culture that was almost destroyed and had disappeared from the view of its own people and society during the previous regime. Today male and female students studying at the Secondary School of Fine Arts complete a full education that secures the future for their traditional culture. After diligent work for more than two decades, many graduates from the school perpetuate the traditional culture for the next generation. Some of them teach in the schools, and some work in the department of performing arts, while the rest join different groups and arts associations.

Dance training

Although a clear curriculum is adhered to in the school, modern teaching methods are still not highly regarded in a system where traditional methods still prevail and are practiced today. Both the teachers and students adopt the traditional way of learning and transferring knowledge. Every day, students spend their morning practicing and rehearsing with their masters, who hold a stick that they use to beat the music rhythms – sometimes to correct students as well. Training with a high respect for their masters, the students remain silent when they are being corrected and restart their dance sequences as their masters tell them to. They have to accomplish all the steps and movements that their masters show them.

Furthermore, after finishing their school day, students usually have other private classes at their masters' houses. These after-school classes are an important time for students to receive full-time training with their masters, and they are generally held after finishing the morning classes. While the school classes give general training to many students, it is in the private classes with their masters where they are able to improve their technique. Similarly to the students, the teachers practice the same way. Normally, at school, dance masters transfer their knowledge to all the students equally, but they will teach separately someone whom they want to be a principal dancer or someone who wishes to learn specifically from them. These private classes are free, and students do not need to pay anything to their masters, but they usually bring food, fruit, or a gift to their respected masters. The knowledge and experience that each master passes on to their students from generation to generation comes from many years of study, practice, and self-discovery rather than through studying books or other teaching materials. 'Memory based' teaching has definitely been the most important method that the masters apply to their teaching. In this respect, Cambodian dance education is unique.

Dance institutions – a new challenge

Although the country is now peaceful and developing, dance still faces an uncertain future. As the era of new technology arrives, traditional culture is threatened by the globalization of new culture that has a powerful influence on the young, curious, and susceptible generation of the nation. Everyone seems to be concerned, but little effort is being made to address the issue. Although RUFA and the Secondary School of Fine Arts are considered the single most important places for arts training and preservation, they are not in a position to stem the flow of new culture. At present, the Secondary School of Fine Arts is already experiencing difficulties with the number of student enrolments decreasing year by year – it is a critical situation. Coupled with this, the teaching standard is not as it was – shortages of documents, equipment, and teaching recourses are some of the obstacles. Most teachers prefer to teach students through their own experiences rather than through written texts, and this has resulted in some disagreements amongst them. The new school location that is more remote also prevents students from attending classes. The lack of group solidarity is also a school problem. In the dance department, there are professors who are highly trained in technique, but they are not always able to work together to accomplish goals, negotiate solutions, and discuss problems. Low salaries for the teachers and their low status in society also means that they must try to find other work to maintain their income to support themselves – this affects their teaching, the school, and the students and impedes the school's progress, adding to the range of current problems.

Besides these examples, the current scope of the curriculum and limited written documents are concerns. Unaware of the working environment in today's society, the schools do not adequately prepare their students for the challenges of the outside world. Students, especially those in the Faculty of Choreographic Arts and the dance department, find it difficult to find work and the opportunity to earn money to support themselves.

With fewer public performance opportunities, and the lack of support in society, traditional culture, as well as RUFA and the Secondary School of Fine Arts, is faced with the danger of diminished importance. Apart from some grants that are provided by foreign organizations, for example LINC, Arts Network Asia, Amrita Performing Arts, the French Cultural Center, and some sponsors from the government, it seems difficult to find funds in order to support dance works and the dancers. Many are forced to leave their jobs, and students drop out of school because of the uncertain future.

Some of the problems that the schools are currently encountering are complex in terms of a successful 'whole of school' approach where strategic goals, plans, and projects can be cohesively coordinated and implemented. The conservative underpinning of the curriculum is another issue in developing Cambodian dance education. The nostalgic past and the pride for their ancestors prevent Cambodian people, in particular the classical dance masters, to change what they have been taught, seen, and heard. Losing identity and traditional culture is a great fear for them, which makes them hold on very tightly to what they have. Sometimes this precludes the acceptance of other possibilities.

To conclude, because of the recent rapid economic development, many countries seem to lose their sense of self-appreciation and value. In this critical uphill battle to maintain traditional culture in the face of the increasingly homogenous global community, everyone has to undertake the mission of nurturing and preserving the culture to ensure a bright future. The schools, especially, should work extremely hard to control the fast flow of new culture, while a new administrative strategy is urgently needed to sustain the traditional culture. A precise mission, strong commitment, and high solidarity amongst the arts communities will empower the schools to achieve their goals.

It is evident that the schools have worked successfully to produce a new generation of dancers as 'cultural preservers'. Besides this, what they have to focus on is carrying these traditions into the future and preventing dance from becoming obsolete in the modern era. The primary concern is for the schools to keep developing so that they can adapt to modern society – they should pay particular attention to this as they continue to evolve. However, innovation has to take place with respect to the traditions and occur in a context where both the past and the present can co-exist. Naturally, everything has to transform to adapt the evolving changes to be engaged and sustained.

Note: This is a revised and updated version of the original chapter in Burridge, Stephanie and Frumberg, Fred (ed.) (2010) *Beyond the Apsara: Celebrating Dance in Cambodia*, Routledge, India.

Notes

1 This sculpture and others found at the site are from Angkor Borei, Funan (1st–6th C) in the present day Takeo province – this pre-Angkor style is known as Phnom Da. They can be seen at the Cambodian National Museum, Phnom Penh.

2 Maspero, Georges, 1929, *Un Empire Colonial Français: l'Indochine*. Vol. 1, Paris: G. Van Oest.

3 This was the 1970 coup whereby Prince Norodom Sihanouk lost his power as the head of state. Subsequently Prime Minister Lon Nol became the head of state for the new Khmer Republic (République Khmère) government.

China

5

CONCERT DANCE IN CONTEMPORARY CHINA

Crossing borders while maintaining difference

Emily Wilcox

Introduction

If we compare concert dance with haute cuisine, contemporary China would be for dance enthusiasts what France is to food lovers. The sheer long-term investment of human labour, material resources, and financial capital to the development of concert dance in China has made the country a leading force in today's global dance marketplace. An enduring commitment to localized tastes and aesthetic expectations has combined with exceptionally high standards of technical ability and production quality to foster a concert dance scene marked by extreme dynamism and distinctiveness. Much of the dance work that appears on stage in China today is rarely replicated elsewhere, either because of the large scale and unusual technical requirements needed for many of these productions or as a result of their close interweaving with domestic social, political, historical, and cultural issues and audiences that makes them less easily transferrable to other contexts. Much like locally sourced restaurants whose menus and flavours reflect the ecology and tastes of nearby communities, the concert dance scene in contemporary China is rooted in localization to the needs and desires of its unique environment.

To continue the metaphor further, China's concert dance, like French cuisine, requires a seasoned palette and familiarity with the milieu to be fully appreciated. To the virgin diner, a slice of Roquefort cheese with its pockets of blue mould and sweat-like moist surface may seem repugnant, if not inedible. The smell alone, combined with the fact that it is customarily not refrigerated, may leave even the most adventurous eater faint hearted. And yet, to those with a penchant for rare cheeses who have spent time in Southern France, this unusual combination of flavour, sight, and texture conjures the warmest of sensations, offering one of the finest delicacies the experienced mouth can enjoy. Thus, while unappealing to some, for others, this substance, especially when savoured in its ideal environment alongside wine, fruit, and warm summer air, is simply exquisite – a perfect combination of tang, crumble, and creaminess that sends the taste buds singing and inspires cravings for more.

Concert dance in contemporary China possesses a comparable sense of '*terroir*' – a special flavour or quality nurtured over time through intensive cultivation in a specific

DOI: 10.4324/9781003160007-9

environment – and thus offers a distinctive dance experience that, like other art forms with rich cultural histories and devoted circles of patronage and connoisseurship, is tailored to indulge a particularly attuned set of tastes. Concert dance in contemporary China is not equally meaningful or enjoyable to all viewers, nor, I would argue, should it be. Rather, what defines this dance scene as a mature artistic environment is precisely that it has fostered a complex aesthetic culture of its own.

This scene has not developed in isolation from international dialogue but rather has benefited from it. That is, the vibrant local character of concert dance in contemporary China has resulted from an intentional and highly self-reflexive effort to be actively engaged with dance trends happening in other parts of the world without being subsumed into any singular hegemonic global dance aesthetic. Concert dance in contemporary China, by virtue of its diverse array of distinct and well-supported dance genres, its long history of engaging in international exchange while emphasizing local difference, and its dense institutional structures for continually developing new domestic talent and choreographic innovation, has evolved into one of the world's great destinations for theatrical dance. In contemporary China, audiences enjoy exceptional experiences of dance for the stage that are at once in tune with international developments and rooted enough in local history and culture to be truly distinctive.

Diversity of genres

Any dance enthusiast spending time in China is likely to be shocked by the large number of concert dance performances happening throughout the year. Beijing, as the national capital and home to many of the country's top dance conservatories and leading dance ensembles, has the most active and wide-ranging dance performance calendar. Major metropolitan regions and provincial capitals all over China also have state-of-the-art theatres as well as smaller venues with active concert dance programmes. Regular dance festivals and national, regional, and local dance competitions ensure a constant stream of new choreography. Meanwhile, nearly every one of the hundreds of professional state-sponsored dance ensembles and university dance programmes has its own home theatre where it frequently stages performances.

A wide variety of dance genres appears in these productions. Some of them, such as ballet and modern dance, are relatively familiar to international audiences. Others, such as military dance, Chinese classical dance, and Chinese national folk dance, are more specific to China. Regardless of genre, however, all domestic concert dance productions are rooted in some way in the local environment and its institutional and cultural ecologies. This may be in how dancers' bodies are moulded by China's unique system of professional conservatory dance education, including its practices of recruitment, tracks of specialization, and common career pathways and lifecycles. It may also be in how different genres of dance have historically been funded, created, and programmed in China. This may include cultivated expectations about what productions in different dance genres ought to look like and how collaboration, remuneration, and authorship operate within creative teams, as well as what the ideal links are between dance choreography and local historical, cultural, and political material. While the genres of concert dance in contemporary China are constantly in flux and often blur into one another, their distinct qualities and trajectories can be mapped and identified over time. At base, the maintenance of this genre diversity is itself a fundamental feature of concert dance in contemporary China.

Military dance

Among the many genres of concert dance regularly performed on China's stages, one of the least likely to be found anywhere else is military dance (*junlü wudao* 军旅舞蹈), also known in Chinese, somewhat confusingly for those outside China, as contemporary dance

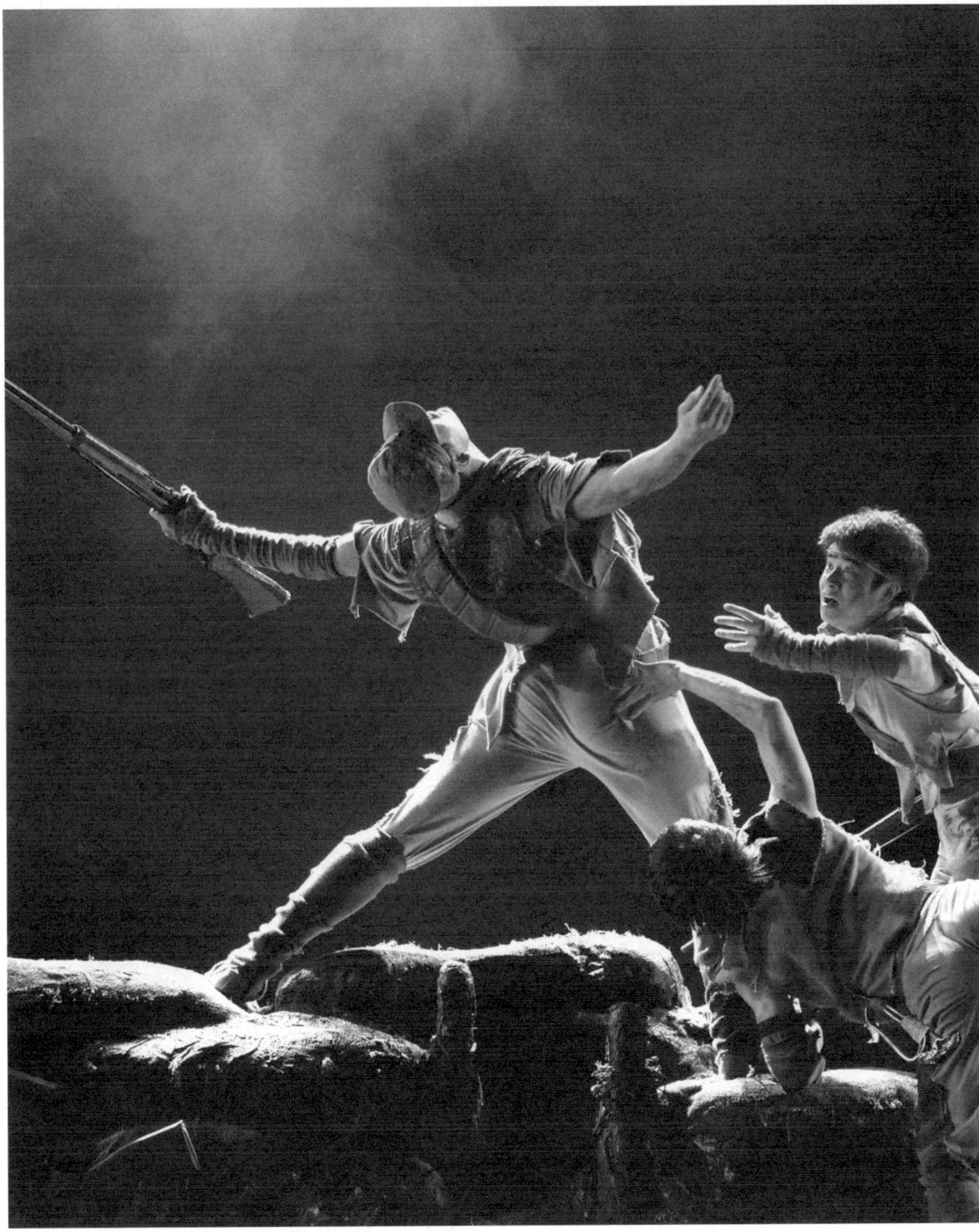

FIGURE 5.1 *Our Father's Generation* 父辈

Source: Performed by Nanjing Military District Political Department Front Line Cultural Work Troupe, ca. 2009

Photographer: Ye Jin (used with permission)

(*dangdai wu* 当代舞). Military dance has its roots in cultural mobilization troupes of the War of Resistance Against Japan (1937–1945) and the Chinese Civil War (1946–1949). At that time, dance was widely used, sometimes performed at the front lines of battle, to invigorate soldiers and civilians to support the war efforts. Beginning in the 1950s, many of these wartime performance troupes were converted into professional ensembles with support from local, provincial, regional, and national governments (Liu 2011). The most famous of these is the Chinese People's Liberation Army General Political Department Song and Dance Troupe, commonly known as the General Political Department Song and Dance Troupe (*Zongzheng gewutuan* 总政歌舞团) founded in 1953, which was renamed the Central Military Commission Political Work Department Song and Dance Troupe in 2016.

A typical military dance choreography features groups of exceptionally lithe male or female dancers in form-fitting army green costumes resembling military uniforms. Often, the look is completed with belts, boots, gold buttons, ammunition packs, water flasks, berets, and so on. Depending on the choreography, dancers may perform with rifles or other weapons as props. If a story is set during the Civil War, dancers will be dressed in the grey-blue uniforms with red-banded lapels and red-starred caps of early Chinese Communist Party soldiers. Similar to war films, these dances are emotional rollercoasters. They tend to begin with tense, stealth movements that build up to extreme high-energy action sequences, typically set to heart-pounding marches with horns and flashing lights and filled with bursts of flying leaps, flips, and other acrobatic feats. This is intended to simulate the life-and-death struggle of active combat through live performance. Inevitably, some soldiers will be wounded or killed in battle, producing another indispensable element of military dance choreography – the tear-jerking scenes in which comrades-in-arm witness and mourn the loss of their friends.[1]

Heroic poses and melodramatic acting are a hallmark of these dances, which deliver impassioned performances filled with high sentimentalism and themes of patriotism, national history, friendship, intimacy, trauma, and commemoration. These dances are designed to inspire reflection on the bravery and sacrifices made by past generations of Chinese soldiers and revolutionaries, usually with the aim of invoking renewed feelings of commitment to the national cause among contemporary citizens. Apart from being staged at military banquets and military arts events, these performances also often appear in live and televised galas celebrating national holidays and political anniversaries. Significant prestige and benefits are associated with employment in military performance ensembles, so many of China's most talented dance artists have appeared in military dance choreography over the years.

Chinese national folk dance

Another genre of concert dance specific to China is Chinese national folk dance (*Zhongguo minzu minjian wu* 中国民族民间舞), also translated as Chinese ethnic and folk dance or simply Chinese folk dance. Like military dance, the concert form of Chinese national folk dance began to develop during the wartime period of the late 1930s and early 1940s and was institutionalized at the national level in the 1950s with support from state-sponsored professional dance schools, ensembles, and cultural bureaus, many of which remain active today (Wilcox 2019). In China, scholars distinguish between Chinese national folk dance, a conservatory concert dance genre, and traditional or popular folk dances that exist outside the conservatory setting. Chinese national folk dance draws its artistic inspiration from folk dances and other local traditions. However, it emphasizes the creative development of these materials into new contemporary choreography adapted for the concert stage (Wilcox 2018). Each style of Chinese national folk dance is associated with a specific region or ethnic group from around China, marking the source from which it draws its inspiration. In this way, Chinese national folk dance brings local performance practices to the concert stage under the rubric of a 'national' dance form, yet it also embodies the regional and ethnic diversity contained within the contemporary Chinese nation-state.

A common misperception about Chinese national folk dance is that it primarily includes dance styles representing ethnic minority groups. Colloquially, the term *minzu wu* (民族舞), literally meaning 'ethnic dance' or 'national dance', is often used as a synonym for Chinese national folk dance. Because *minzu wu* is frequently equated in popular usage with ethnic minority dance, this usage obscures the important role that dances of the Han majority also play in this genre. The fact that many non-experts cannot distinguish between Han and non-Han dance and mistakenly believe that all 'colorful folk dances' are ethnic minority dances further exacerbates this misunderstanding (Wilcox 2016). In fact, some of the earliest and most influential works of Chinese national folk dance were stage adaptations inspired by *yangge* 秧歌, a rural Han folk form from northern China (Wilcox 2020). Today, Han folk dance styles, many derived from regional *yangge* forms, remain a vital component of Chinese national folk dance. One such form is Dongbei (Northeast) Yangge (东北秧歌), in which dancers create a playful and celebratory mood by tossing and spinning bright red eight-point star handkerchiefs as they cross their feet over one another, bob their shoulders up and down, and swing their chin and hips from side to side.[2] Another example is Shandong Jiaozhou Yangge (山东胶州秧歌), in which dancers (usually women), perform graceful twisting actions in the knees and torso and carve smooth, swerving figure-eight patterns through the air with their arms and wrists while holding a large silk fan in one hand and a gauze handkerchief in the other.[3]

Ethnic minority dance also makes up an important component of Chinese national folk dance. Among the most widely performed are Uyghur, Tibetan, Mongolian, Korean, and Dai dance, although there are now dance styles and repertoires representing nearly all of China's 55 recognized minority groups. Each form of Chinese ethnic minority dance has its own distinct techniques and repertoires, and these are often further broken down into numerous sub-styles, most of which also have masculine and feminine versions that can be quite different from one another. Because of the enormous variation both between and within minority dance styles, it is difficult to describe Chinese ethnic minority dance as a whole. However, Uyghur dance is generally known for its percussive spins and expressive hand gestures, Tibetan dance for its rhythmic stomping footwork, Mongolian dance for its swaggering machismo and shoulder isolations, Korean dance for its combination of floating arms and sinking pelvis, and Dai dance for its flexed wrists and slithering torso. Like Han folk dance, ethnic minority dance is also constantly evolving, with each generation of choreographers developing their own creative interpretations of existing styles and themes while also discovering new sources of inspiration.[4]

Chinese national folk dance is part of the training of nearly every professional dancer in China, regardless of discipline. Its works are staged in every kind of performance venue, from the most ornate of grand national opera houses to the humblest of remote tourist sites and school gymnasiums. Chinese national folk dance is a staple of most dance competitions,

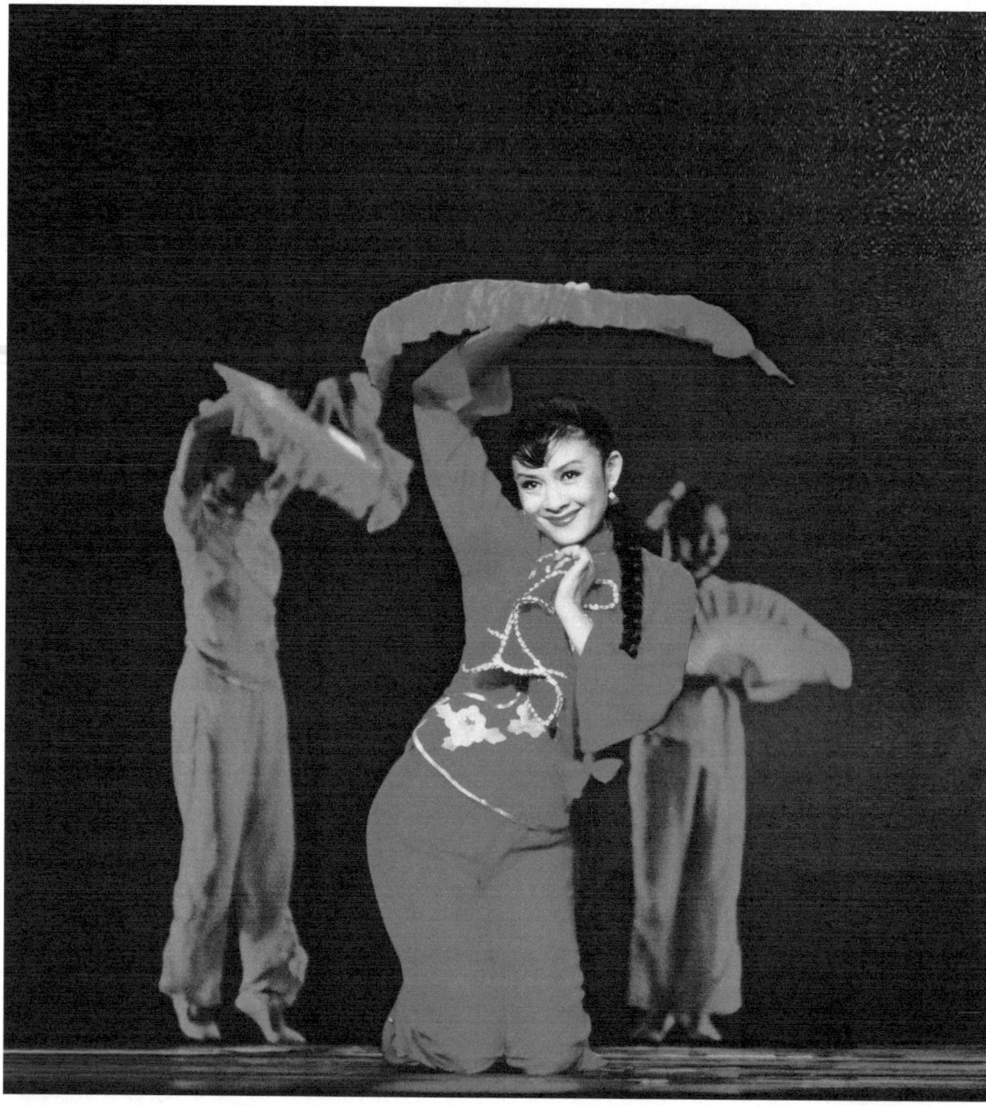

FIGURE 5.2 Chinese national folk dance *yangge*-inspired choreography *Spring Wind and Willow* 春风杨柳

Source: Performed by Jilin Provincial Song and Dance Theater, ca. 2001

Photographer: Ye Jin (used with permission)

performance festivals, and entertainment galas, which are presented live as well as on many major television stations. The popularity of this dance genre is perhaps most evident in public urban life, where amateur enthusiasts learn and perform it daily in parks and squares all over China. Chinese national folk dance is also a common symbol of Chinese culture internationally and is practiced among Chinese diaspora communities and others interested in Chinese arts worldwide.

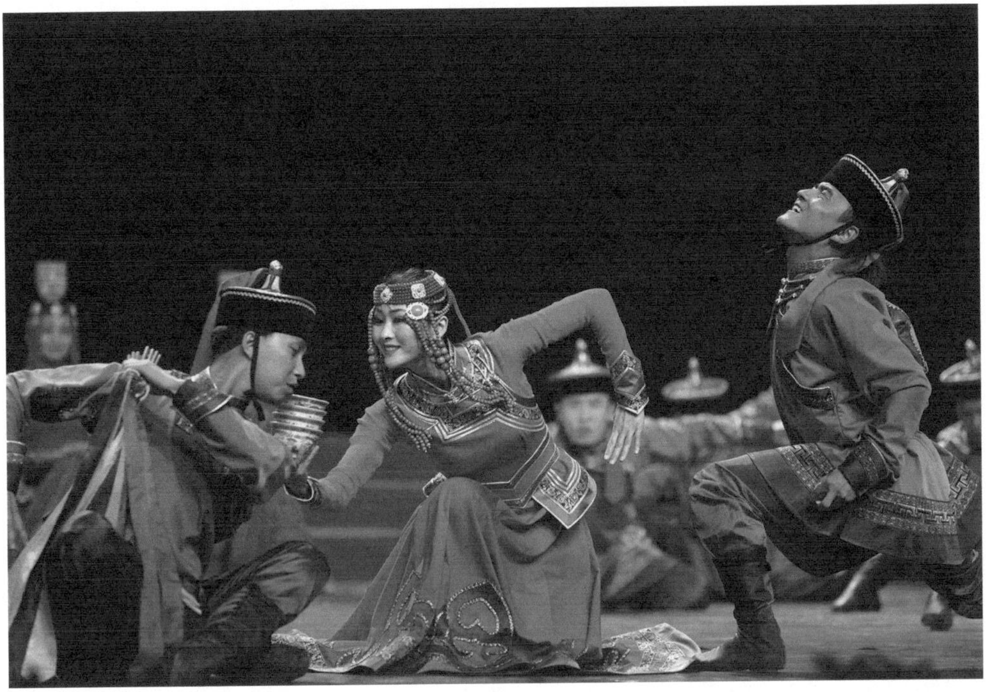

FIGURE 5.3 Chinese national folk dance Mongolian-style choreography *Cup Bowl Chopsticks* 盅碗筷

Source: Performed by Inner Mongolia Ethnic Song and Dance Theater, ca. 2011

Photographer: Ye Jin (used with permission)

Chinese classical dance

Chinese classical dance (*Zhongguo gudian wu* 中国古典舞) is the third major genre of concert dance specific to China. It first developed in the 1950s through collaborations between dancers (one of whom was famously from Korea) and Chinese theatre actors, especially performers of Peking opera (*Jingju* 京剧) and Kun opera (*Kunju* 昆剧, also known as Kunqu 昆曲). Despite being a latecomer to the scene, Chinese classical dance quickly rose to prominence as one of China's most important national dance forms. By the late 1950s, it enjoyed a status similar to military dance and Chinese national folk dance in terms of the attention it received within the professional dance field, as well as recognition in the national media and government support (Wilcox 2019). Chinese classical dance was created to serve as the movement language for Chinese-style long-form narrative dance choreography, known in Chinese as 'national dance drama' (*minzu wuju* 民族舞剧), which was first developed in the late 1950s. National dance dramas composed using Chinese classical dance continue to be extremely popular in China today, where they make up a major category of new concert dance choreography. The leading ensemble creating works of this kind is the China National Opera and Dance Drama Theater (*Zhongguo geju wujuyuan* 中国歌剧舞剧院, est. 1953), whose recent productions include the Chinese classical dance national dance dramas

Confucius (*Kongzi* 孔子 2013), *Princess Zhaojun* (*Zhaojun chusai* 昭君出塞 2016), and *Li Bai* (李白 2017).[5]

The term 'classical' in Chinese classical dance has often been a source of confusion, both in China and abroad. Some mistakenly confuse Chinese classical dance with Chinese ballet because 'classical dance' can refer to ballet in both English and Chinese. Others wrongfully believe Chinese classical dance is a directly inherited ancient court or religious tradition, the way the term 'classical dance' is often used in other parts of Asia. There are some elements of truth in both of these interpretations. The early creators of Chinese classical dance did borrow from Soviet ballet in their quest to create a Chinese national dance form that could be used to train professional dancers and convey full-length narrative stories. Also, as Chinese classical dance has developed over the decades, new branches of the genre have been added that take increasing inspiration from historical materials (Wilcox 2012a, 2019; Jiang 2020). Nevertheless, both of these interpretations are more problematic than helpful overall. This is because Chinese classical dance is an entirely different dance genre from ballet, with its own distinct movement techniques, educational systems, aesthetic expectations, practitioners, and repertoires. Moreover, Chinese classical dance is not an ancient dance form. It is a mode of contemporary concert dance that derives inspiration from traditional aesthetics and sources. The genre's creativity and artistry lie in its choreographers' efforts to imagine and produce impressions of Chinese classical culture from within a contemporary art form.

Movement elements that are often found in Chinese classical dance choreography include inertia, fluidity, and the appearance of circulating breath, particularly in the upper body and chest but also in the arms and legs, especially the joints. This appearance of breath may be expressed through sudden rising and falling actions; circling and spiralling movements that run parallel, vertical, or diagonal to the floor; and the unfurling of full-body extensions that stretch and flex through space rather than holding fixed lines. As in military dance and Chinese national folk dance, the entire body is theatrically expressive in Chinese classical dance, including the dancer's face and eyes, as well as her fingers and even toes. Props are also common in Chinese classical dance. Whereas military dance often employs guns, and Chinese national folk dance features fans, handkerchiefs, porcelain bowls, and so on, in Chinese classical dance, the most common props are metal swords and long silk sleeves. These objects amplify the dancer's physical movement and add technical difficulty to the choreography. At the same time, they also provide mediums of emotional expressivity, vectors for aesthetic experimentation, and cultural symbols related to the characters or themes being explored.

Ballet and modern dance

Ballet (*balei wu* 芭蕾舞) and modern dance (*xiandai wu* 现代舞) are also important components of the concert dance scene in contemporary China. China has several professional ballet companies, the oldest and most famous of which is the National Ballet of China (*Zhongyang balei wutuan* 中央芭蕾舞团), established in 1959. China's ballet ensembles perform both European classics such as *Giselle* and *Swan Lake* and newly created Chinese-themed ballet productions. The latter are often based on Chinese theatrical or literary works, events from Chinese history, or well-known Chinese films. While ballet productions on Chinese themes often include Chinese aesthetic elements

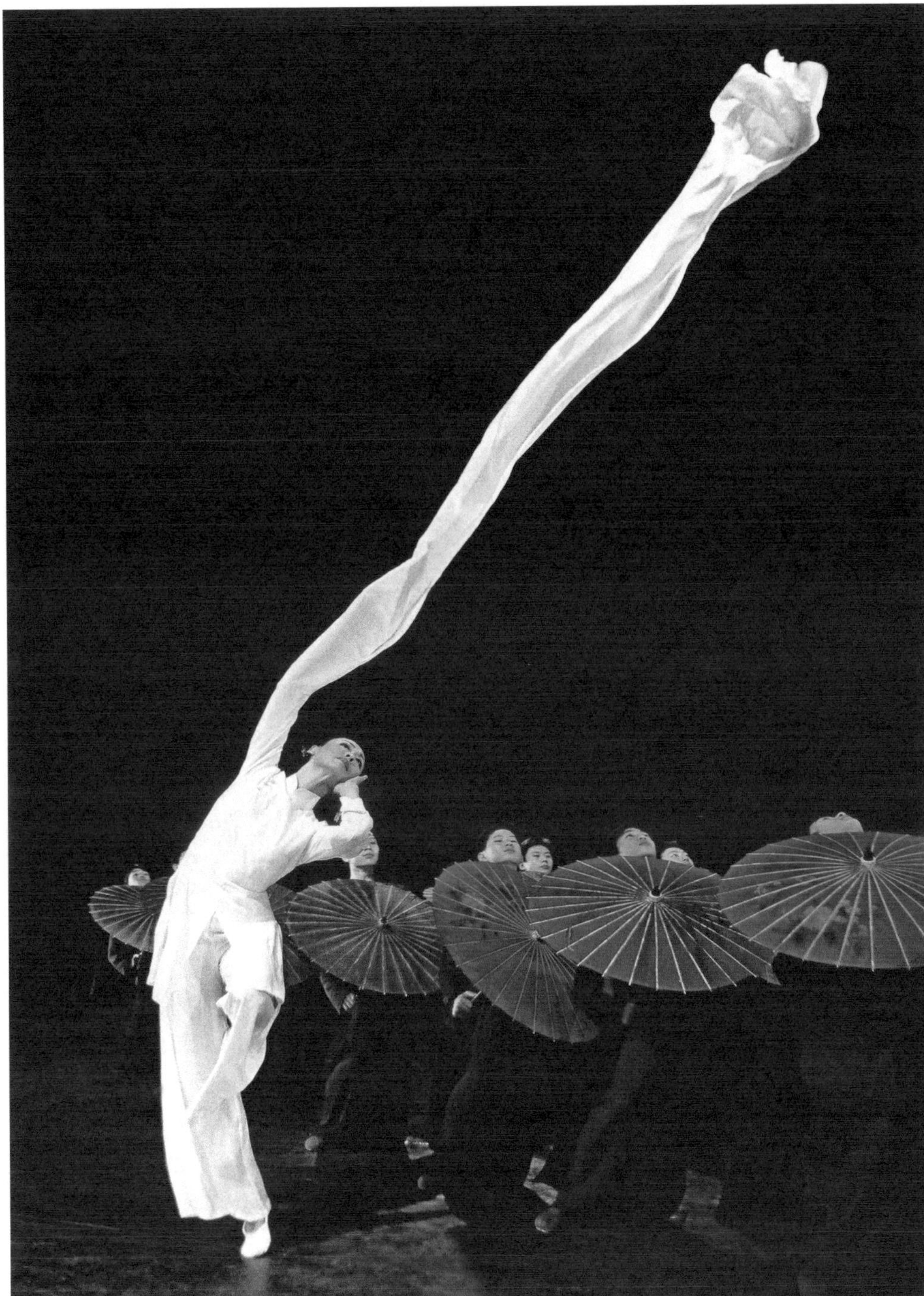

FIGURE 5.4 Chinese classical dance choreography *The Mother River* 黄河母亲

Source: Performed by People's Liberation Army Art Academy, ca. 2001

Photographer: Ye Jin (used with permission)

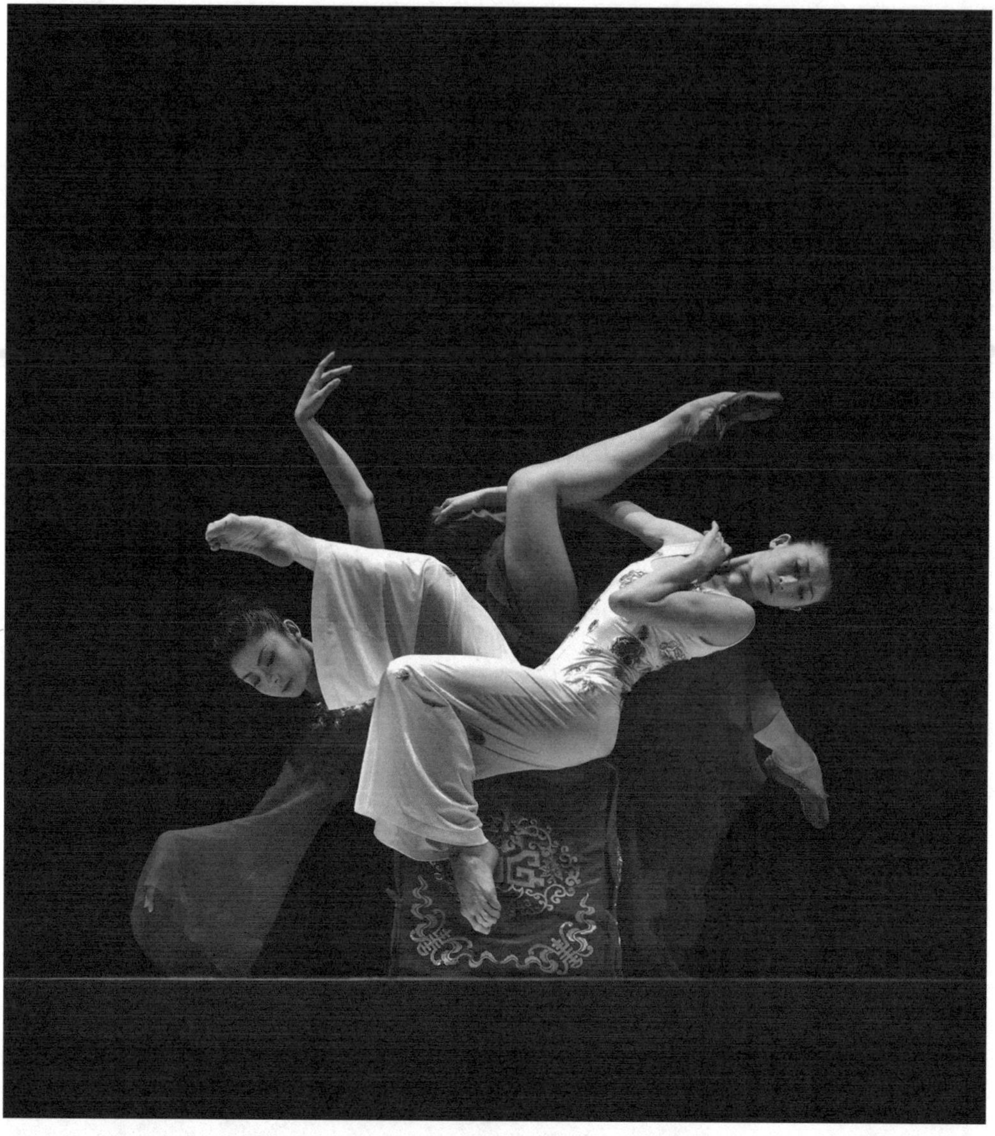

FIGURE 5.5 Chinese-themed ballet *Peony Pavilion* 牡丹亭

Source: Performed by National Ballet of China, ca. 2008

Photographer: Ye Jin (used with permission)

in the sound scores, costuming, and set design, the movement vocabularies and choreographic conventions in these productions are typically drawn from ballet, making these works choreographically distinct from national dance dramas based on Chinese classical dance or Chinese national folk dance.[6] Female dancers in Chinese-themed ballet productions usually perform on pointe, and costumes are often cut to reveal the dancers' shoulders, upper arms, sternum, and legs, which differs from costume designs usually used in military dance, Chinese classical dance, or Chinese national folk dance. New works developed by Chinese ballet companies often incorporate movement and choreographic approaches from modern dance, bringing them in line with trends in ballet companies worldwide. As a whole, ballet operates as a transplanted 'foreign' genre in contemporary China that draws value from its association with elite European aesthetics and culture. Ballet companies not only perform works from classical European (predominantly Russian and French) ballet repertoires, they also participate in the Eurocentric international ballet community and produce new choreography that largely aligns with the tastes and expectations of this field.

Modern dance is one of contemporary China's most rapidly growing genres of concert dance. The first professional companies in China specializing in modern dance were established in the 1990s, beginning with the Guangdong Modern Dance Company (*Guangdong xiandai wutuan* 广东现代舞团) in 1992 and followed soon after by Wen Hui's Living Dance Studio (*Shenghuo wudao gongzuoshi* 生活舞蹈工作室) in Beijing and Jin Xing Dance Theatre (*Shanghai Jin Xing wudaotuan* 上海金星舞蹈团) in Shanghai. Today, China's leading modern dance company is TAO Dance Theater (*Tao shenti juchang* 陶身体剧场), established in Beijing in 2008. Modern dance is unlike contemporary China's other major concert dance genres in that there are few professional conservatory programmes focused on training dancers in modern dance. Rather, most practitioners in this genre train and in some cases perform professionally in other genres before they join modern dance companies. In the case of TAO Dance Theater, for example, the company's founding members Tao Ye 陶冶, Duan Ni 段妮, and Wang Hao 王好 previously trained or performed in military dance, Chinese classical dance, Chinese national folk dance, and ballet before they began practicing modern dance. Unlike many other concert dance genres in China, modern dance practitioners also often gain significant professional experience abroad before setting up their own companies. Duan Ni, for instance, danced with New York-based Shen Wei Dance Arts and London-based Akram Khan Company before becoming a founding member of TAO Dance Theater (Wilcox 2012b). This rich breadth of experience makes modern dance in China particularly dynamic and has led companies like TAO Dance Theater to be recognized as drivers of modern dance innovation globally (see cover photo).

Ballet and modern dance are both commonly categorized in Chinese dance discourse as 'Western dance' (*Xifang wudao* 西方舞蹈 or *Xiyang wudao* 西洋舞蹈). This is because both genres originally developed in Europe and North America and were introduced to China in the early 20th century by artists who had either trained in Europe and North America or had teachers who were trained there (Ma 2016; Wilcox 2019). Despite these associations, however, ballet and modern dance have both become deeply ingrained in the concert dance scene of contemporary China and have developed into local genres in their own right. Many of the artists who led the early development of military dance, Chinese national folk dance, and Chinese classical dance were also trained in ballet and modern dance, so the origins of these fields are mutually intersecting. Chinese-themed ballet productions and new forms of modern dance developed by Chinese artists with diverse dance backgrounds both constitute unique dance developments specific to China. It is the balance and intermingling of these five major genres together as equal players in the scene that make concert dance in contemporary China a vibrant and unique place for dance creation and appreciation.

Structures of support

Since its emergence as a new artistic practice in China during the early decades of the 20th century, concert dance has grown into one of the country's most ubiquitous and flourishing cultural industries, gaining large audiences and fans domestically and internationally and establishing China as a powerful player in global dance culture worldwide. In the 21st century, China is now one of the largest producers of concert dance in the world, and it is also a major international exporter of dance talent, including performers, choreographers, designers, scholars, teachers, entrepreneurs, and producers. China boasts hundreds of professional dance schools and university dance departments, and professional dance companies that enjoy robust support from local, regional, and national governments, as well as private and corporate donors, exist all across the country. Each of these groups produces a significant body of new concert dance works every year, creating new choreography that is constantly being regenerated and reconfigured (Ou 2016).

One factor that has been essential to China's success in developing concert dance over the past century is the country's rigorous and diverse system of professional dance education. Since the first full-time, state-sponsored professional dance conservatories were established in China in the 1950s, Chinese dance students have been recruited between the ages of 9 and 12 (usually following the completion of elementary school) to engage in intensive bodily training in preparation for careers as dancers (Wilcox 2011). This educational method, originally modelled on Russian and Soviet practices, is extraordinarily competitive and places supreme importance on physical ability. Historically, tuition and living expenses at professional dance schools were paid for by the state, meaning that promising recruits could enjoy top-level training regardless of their family's wealth. Recruitment for top schools was also historically done nation-wide, so that children with exceptional promise do not need to grow up in major cultural centres to have access to the highest levels of dance education. Over the years, the establishment of private dance schools and an increase in educational competition overall has made it more difficult for students from disadvantaged backgrounds to be recruited into top dance programmes. Nevertheless, the general structure of professional dance education in China has not changed, and dancers in contemporary China continue to be recruited from all over the country to begin intensive full-time training programmes in dance starting in middle school.

Another factor that has contributed to the successful development of concert dance in China is the government's robust financial support for choreographic research and innovation. During the era of complete state support for professional dance institutions, which lasted from the 1950s to approximately the early 2000s, dance schools and professional companies received funds to pay permanent full-time salaries to dancers and other employees. Housing and other social services were also provided under this system, as were funds to regularly create new dance productions. Starting in the 1980s, commercial dance performances were increased, and many companies began to engage in commercial projects in addition to their state-funded work. Some dance companies, especially those practicing modern dance, also relied on support from private international investors and arts and culture organizations abroad. In the early 2000s, the system of permanent employment contracts for dancers largely came to an end, and most professional dance companies were required to move from full state funding to partially or fully commercial operations. In the 2010s, a new system for arts grants was introduced, and professional companies and schools now apply and compete with one another for large sums of money to produce new concert dance choreography. In addition to funding for new choreography, there are also grants available to support projects in dance education and scholarship. This helps to sustain a large and thriving industry of curricular development and publishing in all fields of dance research, including pedagogy, choreography, history, anthropology, and criticism.

A final key factor that has nourished the dynamism of concert dance in contemporary China is a sustained investment in intercultural, cross-disciplinary, and transnational exchanges and collaborations. All five major genres of contemporary Chinese concert dance discussed previously were established through border-crossing activities of some kind. Military dance, for example, grew out of the introduction of early Euro-American and Japanese modern dance, Western realist theatre, Soviet agitprop, European military marches, and Chinese wartime mass media (Ma 2016). Chinese national folk dance emerged out of collaborations across class (between urban intellectuals and rural farmers, for example), as well as between artists from different ethnic communities within China. Some of the creators of early Chinese national folk dance were diasporic figures trained abroad (Wilcox 2019). As mentioned previously, early Chinese classical dance benefited from the involvement of Chinese theatre practitioners, as well as artists from Korea, and it also took inspiration from Soviet ballet, as well as classical dance forms from other parts of Asia. Over the years, experts in Chinese martial arts, archaeology, art history, and court ritual have further contributed to the development of Chinese classical dance.

In the case of ballet and modern dance, international collaboration has been central to these fields throughout their development. Russian émigrés first brought ballet to China in the 1920s, and over the years, numerous waves of international ballet instructors from different countries have taught and worked with Chinese ballet companies. China's earliest practitioners of modern dance were trained in Paris, Tokyo, and New York, and the establishment of China's first modern dance company in 1992 grew out of visits in the early 1980s by Asian American dancers and, beginning in 1987, cooperation with the American Dance Festival (Solomon and Solomon 1995). Dancers, choreographers, and dance scholars from all over the world frequently perform and teach in China, and their Chinese counterparts also regularly perform, work, teach, and study abroad. This commitment to movement across borders has increased over time, and it has been further enhanced by new data-sharing technologies that facilitate virtual communication even when physical travel is not possible.

Conclusion

Like other countries and regions across Asia and the Pacific, China has managed both to import and localize diverse dance genres from abroad and to create new dance genres out of its own unique cultural materials and historical circumstances. Rather than remaining isolated from one another, practitioners of imported and locally developed dance styles have continuously crossed borders and broken boundaries between dance forms even as they have remained fervently committed to the promotion of stylistic and thematic diversity. This spirit of balancing collaboration with local specificity has allowed dancers in China to continuously forge new experiments and innovations without succumbing to any single universalizing or hegemonic national or global contemporary dance aesthetic.

One way that contemporary China has maintained this extensive diversity of concert dance practices is through the way concert dance itself is theorized and defined in Chinese dance discourse. In much of the English-speaking world, 'concert dance' is an exclusive term reserved for a relatively small range of dance genres and practices, often those that conform to aesthetics and norms defined by ballet and Euro-American modern, postmodern, and contemporary dance. University dance programmes, professional grant funding, and performance programming likewise tend to privilege this narrow range of forms and require practitioners of other genres to assimilate in order to be taken seriously (Chatterjea 2013).

In China, however, the term 'dance art' (*wudao yishu* 舞蹈艺术), also known as 'art dance' (*yishu wudao* 艺术舞蹈), does not discriminate by genre or aesthetic approach. As defined in the *Comprehensive Dictionary of Chinese Dance*, dance art is

> a kind of integrated performing art form that takes the human body itself as creative subject, employing the principle methods of posture, shaping, movement, and technique in close integration with the artistic methods of music, literature (e.g., poetic sentiment, composition, narrative plot), costume, makeup, props, scenery, installations, lighting, and effects, also known as art dance.
>
> *(Wang et al. 2009, p. 568)*[7]

Additionally, the *Dictionary* specifies that

> In terms of expressive traits, art dance can be divided into lyrical or expressive dance (dance of feelings), narrative dance (dance of plot), and theatrical dance (dance drama); in terms of expressive form, [art dance] can be divided into solo, duet, trio, group dance, suite, dance poem, song and dance, music and dance historical epic, etc. In terms of expressive style, [art dance] can be divided into Classical Dance (*gudianwu* 古典舞), Folk Dance (*minjianwu* 民间舞), Modern Dance (*xiandaiwu* 现代舞), etc.
>
> *(Wang et al. 2009, pp. 546–547)*

From this definition, it is clear that China's concept of dance art or art dance is broader and more inclusive than the English concept of concert dance. This inclusivity allows radically different dance genres and their diverse choreographic aesthetics and approaches to coexist within the world of professional dance schools and ensembles in China without one approach gaining dominance over the others. Rather than setting up concert dance as an exclusive category defined by one or two privileged models that others must conform to if they wish to gain access to resources and institutional space, China's definition of dance art or art dance is an open invitation to introduce new forms and develop new possibilities. In other words, it is a definition that recognizes multiple approaches to choreographic innovation as equally valid.

Apart from the five genres discussed previously, there are a wide range of additional dance forms gaining prominence on China's concert stages today, including hip hop or street dance, ballroom dance, jazz, and others. These same forms also have a strong and growing presence in China's dance conservatories. Dance genres from other Asian countries, as well as from Africa and Latin America, in addition to folk dances from Europe, have also at times been taught and performed by China's professional dance schools and ensembles (Wilcox 2017). As concert dance platforms around the world seek to become more inclusive, China offers one model for how diversity can be achieved while cultivating the distinct flavours of one's own *terroir.*

Notes

1 To view *Our Father's Generation*, visit www.bilibili.com/video/BV1oc411h7BA?from=search
2 To view students performing Dongbei Yangge, visit https://v.qq.com/x/page/g0170emm79p.html
3 To view students performing Shandong Jiaozhou Yangge, visit https://www.youtube.com/watch?v=Nr1kak0ZQSg
4 To view *Cup Bowl Chopsticks*, visit www.youtube.com/watch?v=qOx8lIg6Qkw
5 To view these recent productions, visit www.youtube.com/watch?v=HVLt1W7Z6PQ (*Confucius*), www.youtube.com/watch?v=pYYA_k-fIJg (*Princess Zhaojun*) https://youtu.be/xqFGRXliNeY (*Li Bai*)
6 To view *Peony Pavilion*, visit www.youtube.com/watch?v=AzRaCdVPO4I
7 All translations from Chinese are mine.

References

Chatterjea, Ananya. 2013. 'On the Value of Mistranslations and Contaminations: The Category of "Contemporary Choreography" in Asian Dance.' *Dance Research Journal* 45(1): 4–21.

Jiang, Dong. 2020. 'The Dilemma of Chinese Classical Dance: Traditional or Contemporary?' in Katherine Mezur and Emily Wilcox, eds. *Corporeal Politics: Dancing East Asia*, 223–239. Ann Arbor: University of Michigan Press.

Liu, Min 刘敏, ed. 2011. *Zhongguo renmin jiefang jun wudao shi* 中国人民解放军舞蹈史. Beijing: Jiefang jun wenyi chubanshe.

Ma, Nan. 2016. 'Transmediating Kinesthesia: Wu Xiaobang and Modern Dance in China, 1929–1939.' *Modern Chinese Literature and Culture* 28(1): 129–173.

Ou, Jianping 欧建平, ed. 2016. '2015 niandu Zhongguo wudao fazhan yanjiu baogao' 2015 年度中国舞蹈发展研究报告. *Wudao yanjiu* 舞蹈研究 148(January): 1–48.

Solomon, John and Ruth Solomon, eds. 1995. *East Meets West in Dance: Voices in the Cross-Cultural Dialogue*. New York: Harwood Academic Publishers.

Wang, Kefen 王克芬, Liu, Enbo 刘恩伯, Xu, Erchong 徐尔充, Feng, Shuangbai 冯双白, eds. 2009. *Zhongguo wudao da cidian* 中国舞蹈大辞典 [Comprehensive Dictionary of Chinese Dance]. Beijing: Wenhua yishu chubanshe.

Wilcox, Emily. 2011. 'The Dialectics of Virtuosity: Dance in the People's Republic of China, 1949–2009.' PhD diss. University of California, Berkeley.

———. 2012a. 'Han-Tang *Zhongguo Gudianwu* and the Problem of Chineseness in Contemporary Chinese Dance: Sixty Years of Controversy.' *Asian Theatre Journal* 29(1): 206–232.

———. 2012b. '中国的边缘，美国的中心：陶身体剧场在美国舞蹈节 [China's Periphery, America's Center: TAO Dance Theater at the American Dance Festival].' *The Dance Review* 舞蹈评论 2012(1): 59–67.

———. 2016. 'Beyond Internal Orientalism: Dance and Nationality Discourse in the Early People's Republic of China, 1949–1954.' *The Journal of Asian Studies* 75(2): 363–386.

———. 2017. 'Performing Bandung: China's Dance Diplomacy with India, Indonesia, and Burma, 1953–1962.' *Inter-Asia Cultural Studies* 18(4): 518–539.

———. 2018. 'Dynamic Inheritance: Representative Works and the Authoring of Tradition in Chinese Dance.' *Journal of Folklore Research* 55(1): 77–112.

———. 2019. *Revolutionary Bodies: Chinese Dance and the Socialist Legacy*. Oakland, CA: University of California Press.

———. 2020. 'When Folk Dance Was Radical: Cold War *Yangge*, World Youth Festivals, and Overseas Chinese Leftist Culture in the 1950s and 1960s.' *China Perspectives* 120(1): 33–42.

Hong Kong and the region

6

BALANCING ACTS/DECENTRING EXERCISES

The Asia Network for Dance AND+ – a network of its place and time

Anna C.Y. Chan and Angela Conquet (editors), with contributors *Jala Adolphus, Anna C.Y. Chan, Angela Conquet, Kathy Hong, Ophelia Jiadai Huang, Faith Tan*

The Asia Network for Dance, known as AND+, is a young network and the first of its kind in the region. Although the network is still mapping out its modus operandi and collective voice, its ambitious mission, its members' geographical and profile diversity, and the nimbleness of its model are likely to generate much-needed performing arts Asia-specific situated expertise, knowledge, and resources. Inspired by European network models but resolutely and adeptly practising a non-eurocentric approach, AND+ workings and interrogations open a long-overdue space and urgent invitation to consider dance in Asia – in its diversity of practices and on its own terms.

Early days

Similar to most networks, AND+ stemmed from dialogue and conversation between artists and art professionals from several Asian countries, but the context which led to the emergence of such a network and the challenges it strives to address are singular and reach far beyond the Asia region. Officially launched in Hong Kong in May 2018, the network combined four anchoring elements which inherently define its DNA: Asia; Networking; Dance; and Connection, symbolized by the '+' – extended both as an invitation *in* – to what it has to offer now – and *out* – to what may arise in the future as potential for collaboration and change. It is an acknowledgment of the ever-shifting territories and realities dance as an artform is constantly concerned with, arguably more so in Asia than in other parts of the world. The mission of the network is framed around three strategic pillars with a vision to develop, connect, and empower contemporary dance practice by:

- sharing knowledge and information;
- expanding connections for dance professionals, including artists;
- strengthening the visibility of dance within and beyond Asia.

DOI: 10.4324/9781003160007-11

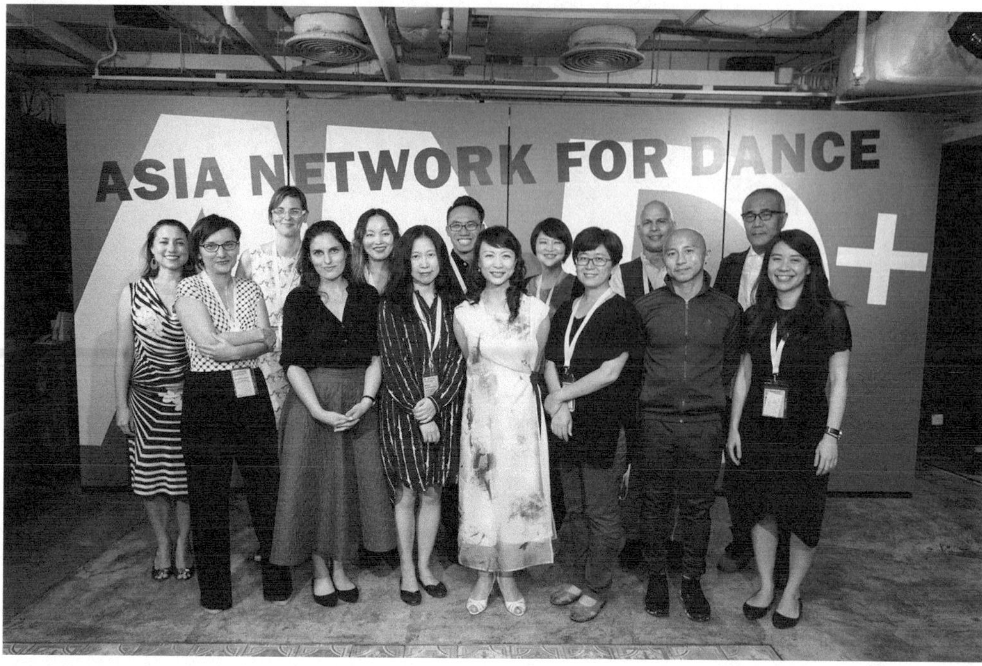

FIGURE 6.1 Fourteen core group members at the AND+ launching ceremony hosted by West Kowloon Cultural District Authority in Hong Kong

Source: Courtesy of West Kowloon Cultural District Authority

The network holds meetings twice a year, hosted by one of the core group members and coinciding with a flagship industry event (Producers' Network Meeting & Forum in Hong Kong, Yokohama Dance Collection and TPAM in Japan, Taiwan Dance Platform, Dance Massive in Australia, Esplanade's *da:ns* Festival in Singapore, etc.) so that the other members have an opportunity to engage with the context and artistic communities of the hosting country.

A core group of 14 members[1] from nine Asia-Pacific countries steer the network. Two radical directions define the organizational distinctiveness of this initiative:

- In order to preserve focus and intimacy of debates, the core group has radically opted to remain closed to other members for an initial period of three years and, as such, operate more like a collective of like-minded individuals, specialists in their areas of expertise, before setting out to refresh the membership. Those who wish to be connected to the network can join its Facebook group. The network benefits from minimal administrative support offered by the West Kowloon Cultural District Authority (WKCDA), Hong Kong.
- Each network meeting has a rotating system of two co-convenors, comprising one co-convenor from the previous meeting and a new one, as a way of not only sharing the workload but democratizing the 'chairing' exercise and preserving the custodianship of the content the meetings generate.

In actual fact, AND+ is both a network of 'people' (such as Aerowaves) and a network of 'venues' (such as European Dancehouse Network), and the diversity of profiles and skill sets gives the tone for current and future achievements of the network and points to the potential to pave the way for new models of generating and circulating knowledge and expertise. The members are all dance specialists in their respective fields and represent a salutary mix of large-scale organizations, smaller venues, artist-run centres, company-run spaces, and tertiary education institutions. The members themselves are executives, artistic directors, producers, curators, and practising artists. The contexts they bring to the table, whether country based, organization specific, or simply experience related, constitute the 'mulch' defining the DNA of the network and the diversity of content and expertise it can offer.

The network evolved somewhat naturally, though most rigorously, out of a series of three meetings held over a year in Hong Kong, Jakarta, and Yokohama, respectively. The initial invitation to organize these working groups was launched by WKCDA and was initially concerned with (co)production and networking strategies for performing arts within Asia, although they were workshopped in dialogue with the European Dancehouse Network (EDN). It is noteworthy that European realities were present in these brainstorming exercises, as the explorations of these initial meetings cemented the singularity of the network's mission and vision and raised important questions for the arts sector in the Asia region: How to extend dialogue and connections beyond Asian borders? How to enable meaningful artistic dialogue, discourse, and exchange within Asia while maintaining an outward approach? What can be learnt from the successes and challenges of other European networks such as EDN, and how would these relate to Asia, if at all? Does Asia actually need a dance network?

A plethora of local artists, producers, and professionals were naturally involved in the conversation, as the meetings were held during widely attended events, thus including a multiplicity of perspectives and profiles from different Asian regions and adding to the initial discussions held principally between Taiwan, Hong Kong, and Singapore venues.

Balancing acts – the context of its emergence

It is extremely important to dwell on the context(s) which prompted the emergence of such a network, as this offers a mirror into what constitutes the remit of this network and an indication of what it might achieve. While there are a few trans-Asian performing arts networks such as Association of Asia Pacific Performing Arts Centres,[2] Asian Producers Platform,[3] Asia Discovers Asia Meeting for Contemporary Performance,[4] World Dance Alliance Asia Pacific,[5] and so on, their small number is largely disproportional to the size of this region and does not do justice to the wealth of dance practices and forms, whether traditional, folkloric, or contemporary, that define the Asian regions. Unsurprisingly, the vast geographical span of Asia, its socio-economic realities, its geo-political forces, the weight of ancient traditions or traditional forms, and the vast array of languages, plus the absence of an overarching policy-making and funding body such as the European Union, make it much harder to develop and sustain networks in Asia than in other parts of the world.

In Asia, more than anywhere else in the world, dance as an artform has multiple embodiments, incorporating without distinction classical, traditional, folkloric, and contemporary forms. It is regulated by cultural policy operating almost exclusively at a national level. Furthermore, the imbalance in funding support to artists and arts professionals with huge disparities between well-subsidized countries and cities – such as Japan, Korea, Hong Kong, Singapore, and Taiwan – and those with almost minimal to no public subsidy system – such

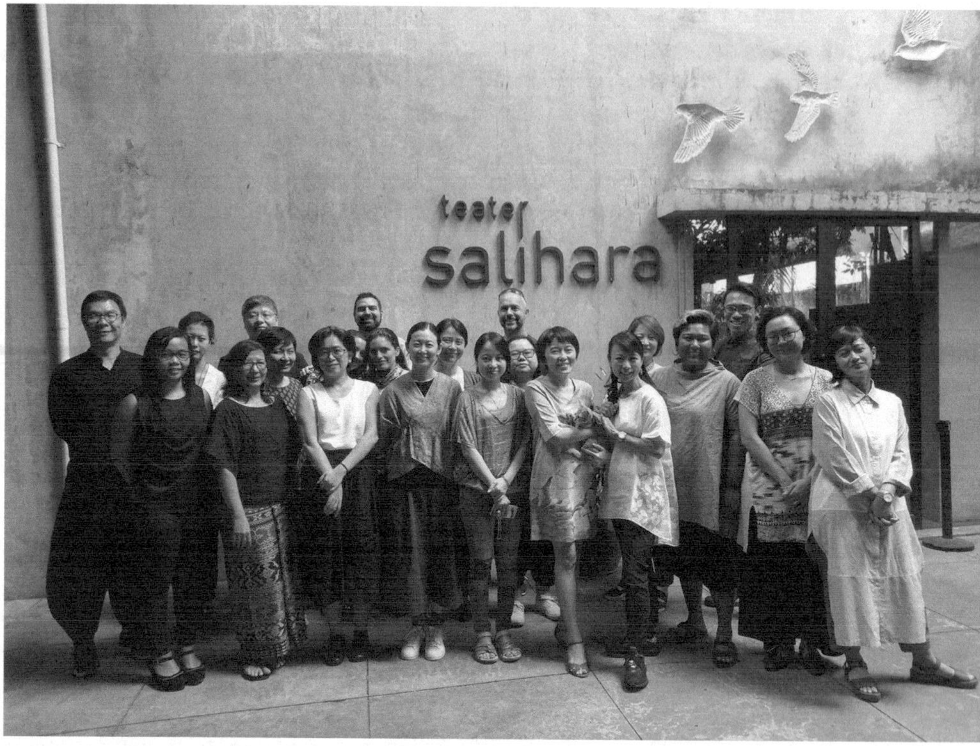

FIGURE 6.2 Participants from around Asia at the second working meeting hosted by Salihara Art Centre in Jakarta, Indonesia, in 2017

Source: Courtesy of West Kowloon Cultural District Authority

as India, Indonesia, and Malaysia – inevitably hinders circulation and production of shared knowledge and of opportunities. It is then by no means surprising to note that until very recently, Asian artists and organizations were more familiar with the European context – through festivals, productions, and collaborations – rather than with their Asian counterparts. Furthermore, some artists have to rely on or even prioritize their European connections and opportunities in order to cement a successful career back in Asia.

It is well proven that international networks, with their multitude of members and vibrancy of meetings, are the fastest vehicle to access local knowledge, opportunities, and expertise. One can attend an edition of Springforward, Aerowaves' annual festival of emerging talent, to get a vivid sense of what predominant aesthetics and formats are currently prevalent in Europe. In Europe, specialized networks grew out of specialized venues and artform-specific policies further strengthened by European funds fundamentally conditioned by trans-national collaboration and mobility.

In Asia, such networks to bring a collective voice and validation were until recently absent or very scarce,[6] and circulation of information, discourse, or simply dialogue was fragmented and usually originating through individuals' projects or connections. It is this precise space that AND+ set out to occupy, within Asia and beyond. The emergence of AND+ was made possible thanks to the need for dialogue and collaboration generated by the new cultural industry realities triggered by the infrastructure boom in certain parts of Asia and simultaneously, the surge of more grassroots practice-based artist-run initiatives looking to exchange around their creative processes outside an outcome-driven economy. Both trends contributed to invigorating Asia-to-Asia exchanges and smoothly steering away from Western-centric models and focus.

In Asia, there are few venues exclusively dedicated to dance, but recent years have seen an injection of substantial funds specifically into large dance/performing arts venues, particularly in China, Hong Kong, Singapore, South Korea, and Taiwan, which cast a spotlight onto these respective regions.

Dating back to the Beijing Olympics in 2008 and the World Expo in 2010, China saw a surge in infrastructure development, starting initially in its megapoles and quickly spreading across the country. In the arts sector, this phenomenon resulted in the construction of new performing arts centres and museums, mostly pharaonic in scale and construction budgets and built by superstar architects. From the surreal Harbin Grand Theatre,[7] inaugurated in 2016 as one of a dozen 'Grand Theatres' built each year in second- and third-tier Chinese cities, to Shanghai International Dance Centre, also opened in 2016, together with recently renovated theatres, new concert halls, a grand opera house (under construction), and a testament to Shanghai's ambition to become the 'capital of performing arts in Asia', there is no doubt that the arts are increasingly recognized as a vital part of a city's economic and social fabric. Moreover, performing arts feature as a Cultural Industry key development area in the '12th Five-Year National Plan' issued by the State Council of China in 2011. According to a recent industry survey,[8] there are already 2,000 theatres in China, mostly built in the last 10–15 years. Such a wave of construction positions 'cultural industries' as a new economic drive and the catalyst to a modern image of an ancient country.

A similar trend of substantial infrastructure investment can also be observed in other parts of Asia. The boost of new arts and cultural infrastructure development in Hong Kong in the last decade is part of a long-term strategy to position Hong Kong as a creative economy and a world city of arts and culture with a strong tie to overall economic impact. An upfront endowment fund of HK$21.6 billion by the government of the Hong Kong SAR to the West Kowloon Cultural District Authority, established in 2008, is meant to develop a reclaimed 40-hectare waterfront site into one of the world's largest cultural hubs, with world-class facilities of performing arts venues, museums, and public spaces. The project includes the purpose-built Xiqu Centre, opened in 2018; the Lyric Theatre Complex; Music Centre; two museums; and an art park. Concurrently, 16 government-run performance venues of varying sizes situated at accessible locations all over Hong Kong territory will have an additional superstructure of the East Kowloon Cultural Centre to be completed by the end of 2020. The revitalization heritage project led by a charitable organization, the Hong Kong Jockey Club, in partnership with the Hong Kong government, opened Taikwun Centre for Heritage and Art in 2018, which offers more venues for performing arts and contemporary arts programming. To capture the long-term economic strategy, many private corporations and developers have also invested in a mix of arts and cultural facilities, invigorating the arts ecology as well as the demand for new artistic content. All these speedy developments foster the growth of cultural industries and performing arts sectors but at the same time prompt many urgent challenges such as programming, audience building, and sustainability issues.

Taiwan's performing arts venues have been, for the past 40 years, largely public facilities built and operated by government entities, such as the National Theater and Concert Hall, situated in Taipei, and the cultural centres existing in each major city. An exception to these publicly run venues were a handful of smaller private spaces, most of them converted into black boxes or studios and dotted across the country. During the last decade, in an attempt to de-centralize cultural resources across the island, Taiwan saw a boom of large theatre structures, including two new national theatres, National Taichung Theater and National Kaohsiung Center for the Arts Weiwuying – the largest performing arts centre ever built – respectively located in the central and southern regions of Taiwan.

The alliances these new and existing large presenting houses sought to form with other Asian colleagues in view of sharing co-production and touring costs, coupled with an evident emphasis on a southeast Asia-bound government policy, prompted the performing arts sector to re-evaluate the until then eurocentric approach that had been the dominating trend in Taiwan and had placed the focus on a Europe-to-Asia circulation of artworks.

Concurrently, Taiwan cultural centres, previously used principally for hire, started to invest in their own curated programmes and productions. In 2000, privately owned small studios and repurposed spaces opened up to embrace live performance. This abundance of new spaces offered a creative alternative to conventional theatre spaces, allowing new formats to sprout, such as the opportunity to create site-specific works. All contributed to a fast-growing performing arts sector. When the Taipei Fringe festival spearheaded the notion of hosting performances in cafes, shopfronts, and galleries, this opened up wide possibilities for makers, space business owners, and audiences which resulted in a vibrant arts ecology no longer dominated by the large players.

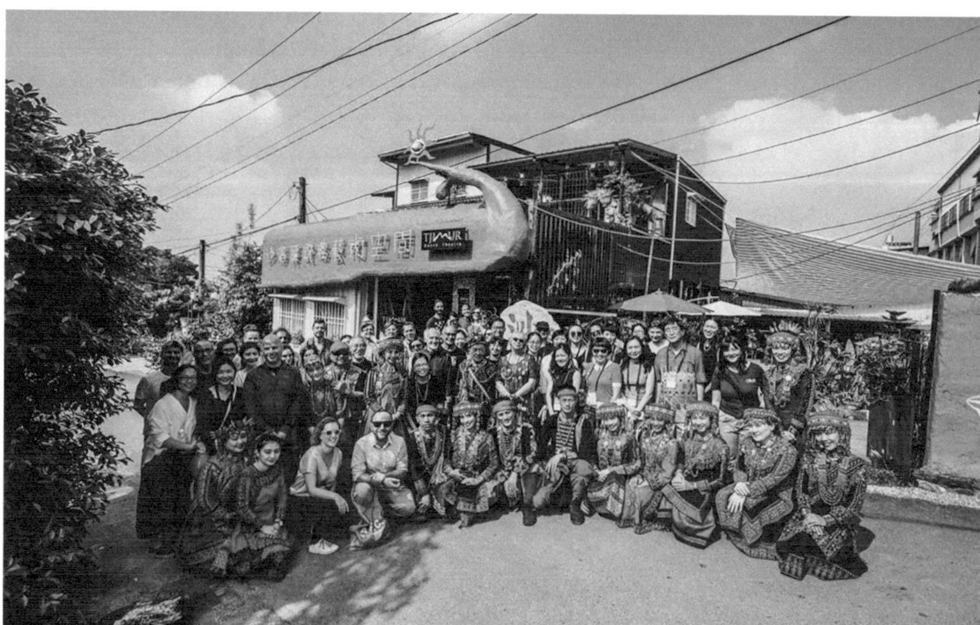

FIGURE 6.3 Visiting Tjimur Dance Theatre as part of the Taiwan Dance Platform hosted by Weiwuying National Kaohsiung

Source: Photographer: LaFun, Courtesy of West Kowloon Cultural District Authority

Another example of massive infrastructure development is the Asian Arts Theatre, part of the Asia Culture Center, a government-funded establishment in the heart of Gwangju City in South Korea. It was regarded as the most ambitious and momentous art project in Korean history in terms of budget, artistic vision, and scale. As then-President of South Korea Park Geun-Hye stated in her inaugural address in February 2013, 'cultural renaissance' was one of her three governmental priorities. Stating its ambition to be the hub for Asian contemporary performing arts, the Asian Arts Theatre's team of local and international curators kicked off in 2015 an international festival with a large programme of commissions and co-productions of Asian and international works. Unfortunately, within only a few years since its inauguration, the Asian Arts Theatre in Gwangju seems to be struggling, with the festival now discontinued.

A resulting phenomenon of the rapid venue development is the imperative need for investment in programming and artistic development. Sustainable development of any performing arts venue requires not only a physical space but, as importantly, stable long-term financial commitment and solid expertise in running these spaces. For example, more than half of the newly built grand theatres in mainland China often remain empty most of the time after their launch because of the substantial operational costs, or they survive mostly via commercial venue rentals to the detriment of any artistic programme.

In reality, lack of consistent public funding support for programming and artistic development is an ongoing challenge for many Asian arts sectors in addressing sustainability.

Programming challenges also come from having to fill huge-capacity theatres – on average 2,000 seats – which allow for a specific kind of programming to attract the masses of audiences necessary to keep these venues economically viable. They are rarely able to support on their large stages more experimental works or smaller-scale projects. More discrete practice-oriented and predominantly artist-initiated or artist-led initiatives started to emerge in recent years, supported by small but nimble organizations such as Dance Nucleus in Singapore or SEA Choreolab Ribun Dahan in Malaysia, working exclusively with independent artists such as Staging Alterity in Shanghai or ADAM in Taiwan.

Decentring exercises

Today, contemporary dance is increasingly present in dedicated organizations, festivals, seasons, residencies, and curriculums all across Asia, but in spite of this recognition, contemporary dance discourses, aesthetics, and production modes (from the pedagogical to the curatorial) remain centred on Western Europe and North American practices. This creates imbalances in authenticity, authorship, and agency and therefore suggests a critical need to advocate for a more global responsibility in nurturing a new discourse and approach for re-locating contemporary dance to its locus of emergence, one which acknowledges a greater constellation of geographical diversity and multiplicity of practices, bodies, temporalities, and cultures.

It is in this very space and landscape that the AND+ situates itself in its current ambitions and endeavours: decentring the hegemony of eurocentric models and reflecting on Asia-specific and Asia-to-Asia realities and modes of production, exchange, and dialogue. Over its previous meetings, the AND+ has outlined three lines of inquiry concerning pressing issues for the development of dance and performance in Asia:

- supporting the development of artistic practice in dance in Asia,
- clarifying expectations for artist residencies, and
- encouraging better guidelines for ethical practices in dance.

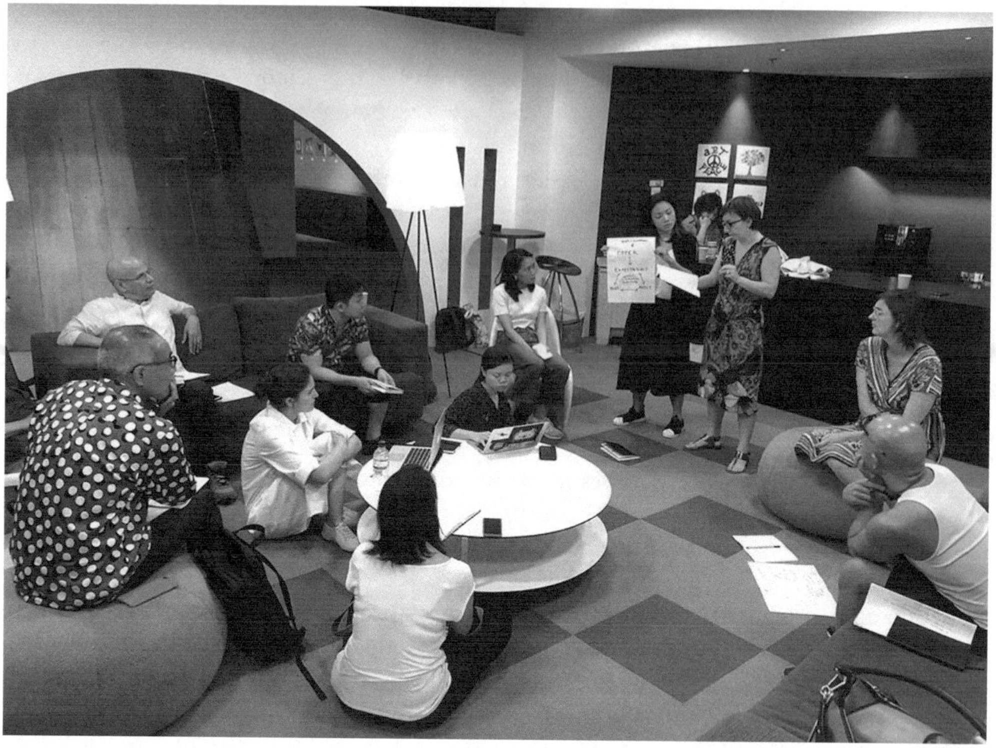

FIGURE 6.4 Core group closed-door meeting and discussion during Singapore meeting hosted by Dance Nucleus and Esplanade

Source: Photographer: Anna C.Y. Chan, Courtesy of West Kowloon Cultural District Authority

The workings led to paying attention to what one means when one speaks of 'Asia', 'practice' and 'contemporary', a grounding exercise that not only highlighted the many subtleties and complexities at play in Asia but, as importantly, the role the network will play in generating 'resource' knowledge within Asia and 'expertise' knowledge between Asia and the rest of the world.

Which Asia?

Although there is a willingness and interest from institutions and platforms both inside and outside of Asia to support and present more dance from Asia, the path towards decentralizing existing eurocentric funding, producing and presenting power systems, barriers, ethics, and habits, requires a deliberate shift to actively integrate and operate with more inclusive mindsets, policies, and processes.

The network sees its responsibility in finding the right tools to decolonize the language used in marketing and discursive exercises; de-bunk 'exoticisation' or instrumentalization of Asian 'other-ness'; and prevent globalization of practice and dominating dynamics of taste, equipping artists to understand and expertly navigate the power of European production modes and the temptation to 'tick boxes' of the system they are part of rather than their own imperatives. Seen from the West, it is often forgotten that there are as many Asias as there are

Asian countries. Its geographic boundaries are vast and its countries diverse, with a diversity of languages, histories, cultures, traditions, religions, and political systems. Europe is bound together – at least in theory – by unifying EU policy guidelines, largely shared currency, many similarities in the infrastructure ecology, and a so-claimed European identity founded on shared values, which ultimately generate the right regulatory systems and resources widely supporting the mobility of artists. In Asia, there is no such thing as one 'Asia'. There are as many Asia(s) as there are Asian countries, and Asians themselves understand their *Asia-ness* as having multiple meanings and interpretations. It is therefore important to acknowledge this diversity and its importance in all explorations or considerations of 'an' Asia.

Dance artists and practices from Asia are therefore to be understood in their own specificities in order to expand narratives and create new frameworks of reception without filtering its aesthetics through eurocentric lenses. AND+ focuses not only on raising awareness of the risk of both exoticization and homogenization of Asian productions which operate within European context promoting 'international' programming but also on creating the context to learn to appreciate and promote Asian dance on its own terms and thus create the circumstances for a greater polyphony of dance in the future.

Dynamics of taste

AND+ discussions have been preoccupied with interrogating, whether informally or in the network's internal workshops, the politics of taste and in particular the politics of arts ecologies that define questions of aesthetic value and the systems that support them. In particular, the network asks what it means to be a practitioner of 'traditional' artforms in a landscape that often privileges 'the contemporary' and how taste can be thought to be hegemonic. Which artforms are seen, funded, showcased, valued, and supported is ever changing and exposes the trends of the time but, more so, the policy and power systems in place.

The word 'contemporary' as applied to 'contemporary' dance has multiple meanings and interpretations in the Asian context. At present, there is a greater awareness amongst dance practitioners of the plurality of contemporary dance artistic practices and histories. Technological advancements have benefited dance through rapid information dissemination and enabled more possibilities for professional knowledge exchange and networking globally. In spite of this recognition, contemporary dance discourses, aesthetics, and operational modes (from the pedagogical to curatorial) remain centred on practices from Western Europe and North America and following a linear trajectory of time and history of dance. This creates imbalances and therefore reflects a critical need to advocate for global responsibility to nurture a new discourse for redefining the notion of 'contemporary' in dance, one which acknowledges a greater constellation in its geographical diversity and multiplicity of practices, bodies, and cultures.

Contemporary dance artists from Asia who work internationally may find themselves in a paradox of having to think globally and act locally. The artist may face a very specific expectation of a contemporaneity, which is not only based on a western-centric normativity but also in a framing of othering. It reflects the problematic hierarchy of global movement languages and the privilege exercised to decide what is 'contemporary' and what can be relegated to the realm of 'traditional'.

Both normativity and othering are an expression of an uneasiness with encountering the unfamiliar, as this means taking a much higher risk. To recognize a dance practice outside of one's own normativity and discourse, perhaps even without a common language might, be uncomfortable but remains necessary. In a lecture performance held at Dance Nucleus during the AND+ meeting in October 2019, contemporary Bharatanatyam dancer Nirmala Sheshandri speaks about disidentification as a choreographic strategy to circumnavigate the landscapes of West-dominated aesthetics.

Dynamics of production modes

This process proves more difficult when underlying systems and structures do not cater to enabling such shifts. For instance, the funding pattern for dance projects or presentations from Asia internationally is either one off or short term, thereby restricting any continued and sustainable development. Within large festivals, the greater the pressure to make dance accessible or to attract large audiences, the larger the risk of a more reductive or exoticizing label imposed on the artist's work. The existing speed and format of international touring schedules (two days to set up and two days to perform) results in artists spending less than a week within one new location, with little ability to grasp who the audiences are and what context the work is experienced in.

AND+ sees its role in inviting non-Asian presenters and programmers to reflect on the curatorial frame which contextualizes an 'Asian' work and the ways in which the work is introduced to audiences. It is encouraging to see that recent round tables and workshops have focused on raising awareness of the need for programmatic processes in Europe to enable more time (and resources) for research to deepen its specificity of knowledge of Asian artists and their work. The presence of AND+ members in this context is vital to generate an informed and expert dialogue.

What practice? – the future projects

Process and criticality

As dance practices in Asia gradually liberate themselves from the focus on theatrical productions to the development of process-oriented or critical artistic practices, AND+ interrogates what the necessary infrastructures and programmes are to support these processes in Asia. In this line of enquiry, the network will reflect on existing structures of education and pedagogy and artist-led and artist-focus initiatives, as well as the shifting role of artists. How can artists and their practices shape the infrastructure, rather than the other way around? This discussion will also consider the question of access for artists from Asian countries where support for learning, exchange and travel is more limited.

Sustainability

Even before pandemic times brought travelling the world to a standstill, the network reflected on our collective role in approaching modes of production, touring, and curating of contemporary dance differently, looking at how a paradigm of being together could be invented in a more sustainable and responsible way. It also reflected on how access to travel leads to the homogenization of narratives and annihilation of cultural difference.

Guidelines for ethical practices

One of the network's areas of action was to propose ethical guidelines of best practice pertaining to conduct in the production of performing arts specifically in Asia, whether in the studio, on stage (copyright and authorship ownership), and in relation to the audience (engagement and communication). Asia lacks a federating advocacy body which could be a go-to resource for artists and creators to understand their rights and responsibilities. Furthermore, dance practice is dominated by very nuanced and specific power relationships inherent to the transmission and dissemination more traditional or classical dance forms operate with and within, often defined by centuries-old patriarchal or guru–disciple hierarchical systems.

The aim is not to provide hard and fast rules to adhere to but rather to provide guidelines and protocols for consideration and adaptation to different cultural contexts and sensitivities, a sort of suggested best code of practice for practitioners, producers, and venues alike and to slowly denormalize these somewhat archaic, abusive methodologies of creating and producing work. The ambition is to invite non-Asian presenters and producers to reflect on their own responsibilities in understanding, identifying, and addressing the ethics of such systems which they may be unwarily encouraging by commissioning a certain type of work.

The intention is also to do a mapping of different organizations that are working on these issues across the region in order to build a resource point and contact list that one can consult, access locally (e.g., whilst touring), or refer to. The AND+'s soon-to-be launched website will be an invaluable concrete asset in this area.

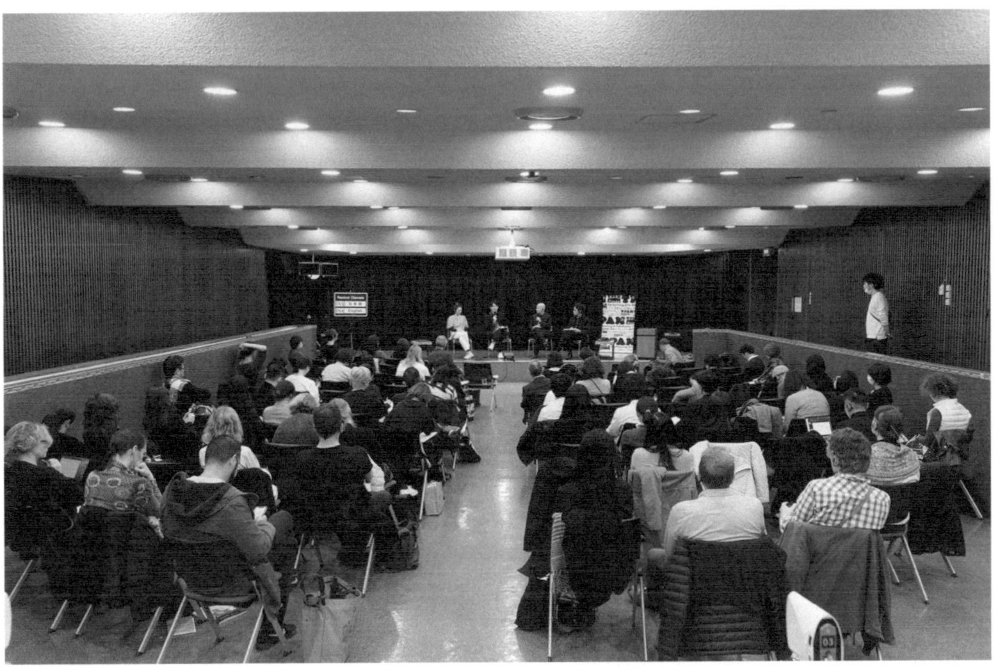

FIGURE 6.5 AND+ plenary session during Yokohama meeting hosted by Yokohama Dance Collection and TPAM

Source: Photographer: Hideto Maezawa, courtesy of Yokohama Dance Collection

Conclusion

AND+ is as such the first dance-specific Asian 'collective' voice for fostering an Asia–Asia dialogue, compiling and connecting existing resources, and acting as a reference point and federating hub for the many and diverse Asian dance communities. But its greater potential and importance may lie in the role it will play in the future in guiding an expertly nuanced and informed understanding of the Asian dance context for those presenters and producers within and beyond Asia who wish to have a grip on the contextual realities and specificities and, as importantly, an Asia-specific frame to understand, appreciate, and engage with the multiplicity of dances coming from Asia.

The network's members have already started essential work on some of the most relevant and urgent issues dance practice and production are faced with in Asia specifically and in the world in general. Through the very frames it offers, AND+ invites reflection on how shifts can be collectively empowered within structures, processes, and mindsets to expand away from biases limiting perception, circulation, and legacy of dance. Raising awareness of the flaws of existing systems is a shared responsibility of the entire dance scene. This role traditionally rested far more on the shoulders of funding bodies, policy-makers, and institution leaders, those who provide the structures where 'international' dance is supported, presented, taught, or recorded.

AND+ has the tools, the expertise, and the agency to enable the emergence of alternative non-normative, non-homogeneous, and de-colonized viewpoints, fighting tokenistic hegemonic approaches and aesthetics and hopefully inspiring a new kind of responsibility. It is in this sense that AND+ is an exemplary model of pan-Asian collaboration, dialogue, and exchange; generator of expertise and situated knowledge; and great activator of relevant interrogations for its European counterparts. It is in its self-assigned role to emphasize the relevancy and interconnectedness of dance beyond geographic boundaries that AND+ is truly a network of its time and place.

Notes

1 List of AND+ members at its launch: Anna C.Y. Chan (Dean, School of Dance, The Hong Kong Academy for Performing Arts; Dance Advisor, West Kowloon Cultural District Authority) Jala Adolphus (Independent Producer, Indonesia), Chen Pin-hsiu (Programming Manager, Cloud Gate Theater, Taiwan), Angela Conquet (CEO and Artistic Director, Dancehouse, Melbourne, Australia), Ahram Gwak (Producer, Team Leader of Programming and Production Team, Korea National Contemporary Dance Company, Korea), Claire Hicks (Director, Critical Path, Australia), Bilqis Hijjas (President, MyDance Alliance; Dance Programme Director, Rimbun Dahan, Malaysia), Kathy Hong (Director of Marketing and Communications, National Performing Arts Center – National Kaohsiung Center for the Arts [Weiwuying], Taiwan), Ophelia Huang (Head, International in Shanghai Dramatic Arts Center [SDAC]; Programme Director, ACT Shanghai International Theatre Festival, China), Daniel Kok (Artistic Director, Dance Nucleus, Singapore), Jayachandran Palazhy (Choreographer and Artistic Director, Attakkalari Centre for Movement Arts [Bangalore], India), Faith Tan (Head of Dance and Theatre Programming, Esplanade, Singapore), Jacky Fung (Manager Programme/China Dance Development, City Contemporary Dance Company CCDC Dance Centre, Hong Kong), and Shinji Ono (Chief Producer, Yokohama Arts Foundation; Director, Yokohama Red Brick Warehouse Number 1, Japan).
2 Website: www.aappac.com/en/home/
3 Website: www.facebook.com/asianproducersplatform/
4 Website: https://adam.tpac-taipei.org
5 Website: www.wda-ap.org

6 A notable example is HOTPOT, a new collective festival within a festival – Yokohama Dance Collection – to promote collectively Japan, Korean, and Hong Kong/Mainland/Macau artists.
7 Designed by well-known Chinese architect Ma Yansong, Harbin Grand Theatre, or Harbin Opera House, situated in China's northern province of Heilongjiang, is a multi-venue performing arts centre measuring 850,349 square feet.
8 Yicai.com, 8 Apr 2019, www.yicai.com/news/100156973.html

India

7

IMAG(IN)ING THE NATION

Uday Shankar's *Kalpana*

Urmimala Sarkar

For a long while, the elite in India did not acknowledge the presence of dance as a part of their culture and existence – then came a stage when dance became *the* emblem of a rich and glorious history and tradition – an image that has stayed. Folk and tribal dances were part of the culture of the unrepresented few, good for showcasing the variety and the 'ethnic-ness' of the Indian people, so they were required to be put in a special category where they were clearly part of the non-elite mass, good for exhibition-like circumstances of the republic day parade or India Festivals abroad but never deemed good enough to be representative of 'high' Indian culture. The 'pure' form of dance came into existence almost through an elaborate engineering process, where the grammar was systematically structured, the link to *Natya Shastra* both deliberately and systematically sought and established, and, in most cases, even the name of the form invented. In this deliberate process of shaping dance history and geography, there was no place for people who did not want to be categorized into either of the two previously mentioned categories: classical and folk.

In the post-Independence era, the dance history of India was narrated with a deliberate political agenda: that of establishing the hegemony of the dominant voice of the nation builders within the nation and building an image of India in the context of the world as a significant south Asian regional entity with a formidable cultural heritage (Menon, 2005: 27–43). More recently, this narrative has been reinforced by the works of government institutions and their funding agencies and archival sources – privileging only those dance forms which can be linked to the 'sanskritized' history. The result is the creation of a museum-like image of a high cultural practice – unchangeable, rare, and exclusive.

In the process of streamlining the heritage of dance in India, the whole question of modernity was subverted time and again. In the re-visiting of dance history, an important task therefore is to first record the multi-layered and multi-phased attempts at modernity by the parallel individual or community/group agencies – as in all other forms of performance – and thereafter set off a discourse on a new historiography of colonial and post-colonial encounters in dance in India. As is explained by Abhijit Pathak in his book *Modernity, Globalization and Identity: Towards a Reflexive Quest* (2006), nationalism remains an important context even in the age of globalization and market economy in India, as the cultural

DOI: 10.4324/9781003160007-13

policies continue to be framed and implemented on the basis of histories and processes formulated in the immediate post-Independence era of planning an ideal image of the nation which was shedding its colonial shackles after a long period of existing as a colonized and oppressed country.

In the context of Indian culture, and especially its dance, it is important to begin an analytical discussion with a discourse on how Indian nationalism has been framed, as Indian dance (even contemporary Indian dance) can only be read within the framework of nationalism in contemporary India. Ernst Renan (1994: 17–21) defines a nation as 'a historically constituted, stable community of people, formed on the basis of a common language, territory, economic life, and psychological make-up manifested in a common culture'. Within the definition itself, there is an acknowledgement of the fact that even if there is an existing group of people in a given territory with a common culture, psychological make-up, history, common language, and economic life, the nation is not a given or an automatic entity but has to be constructed – and acknowledged. Thus, the identity of the nation is something constituted completely through human agency. Renan goes on to say, 'It goes without saying that a nation, like every historical phenomenon, is subject to the law of change, has its history, its beginning and its end' (1994: 20). Hence the idea of nationalism begins from the belief that every nation has a past and must see itself as a territory with a population envisioning a future. Partha Chatterjee (1993: 110–115) writes:

> The idea that 'Indian nationalism' is synonymous with 'Hindu nationalism' is not the vestige of some pre-modern religious conception. It is an entirely modern, rationalist, and historical idea. Like other modern ideologies, it also allows for a central role of the state in the modernization of the society and strongly defends the state's unity and sovereignty. Its appeal is not religious but political. In this sense, the framework of its reasoning is entirely secular. . . .
>
> In fact, the notion of 'Hindu-ness' on this historical conception cannot be, and does not need to be, defined by any religious criteria at all. There are no specific beliefs or practices that characterize this 'Hindu', and the many doctrinal sectarian differences among Hindus are irrelevant to its concept. Indeed, even such anti-Vedic and anti-Brahminical religions as Buddhism and Jainism count here as Hindu. Similarly, people outside the Brahminical and outside caste society are also claimed as Hindu jati.

Talking about the exclusion of the religions like Islam or Christianity, Chatterjee (1993: 110) says,

> Buddhism or Jainism are Hindu because they originate in India, out of debates and critiques that are internal to Hinduism. Islam or Christianity come from outside and are therefore foreign. And 'India' here is the generic entity with fixed territorial definitions that acts as the permanent arena for the history of the jati.

In case of India, for the identity thus carved out of the existing history and played out through its state, directed/patronized cultural practices were the complex narrative of an image – the ideal or the constructed – and hence justified the much-discussed and often-critiqued restructuring process leading to the creation of eight classical dance forms.[1]

Homi Bhaba (1990: 1–7) discusses the narratives of the nations in the introduction of his book *The Nation and Narration*:

> Nations, like narratives, lose their origins in the myths of time and only realize their horizons in their mind's eye. Such an image of the nation – or narration – might seem impossibly romantic and excessively metaphorical.

In this chapter, I take up the case study of *Kalpana*, a 122-minute film that Uday Shankar wrote, directed, and starred in, completed in the year 1948. I look at *Kalpana* as a film which of course was made with a concrete aim to document Uday's creative endeavours in dance but also was a structured, if not cohesive, narrative putting together several important, conflicting, and often contested issues of the anti-colonial, nationalistic, and progressive agenda of that time. The film *Kalpana* remains one of the first and only documents of parallel modernistic endeavour.

In most of the criticisms aimed at his dance, his understanding of Indian culture, and more concretely his film *Kalpana*, Shankar's understanding and efforts have been taken as an individual reaction and reading of Indian culture and society. I would like to suggest through this chapter a new reading of his film and, with the help of that as a concrete document, an understanding of his art. Hence I propose that we place *Kalpana* in that group of films of the post–World War II era of internationalism, whereby the nations were not trying to wipe out their differences and create a 'Global' world of cultural universals but were trying to establish a phase of international exchange of ideas and concepts to strengthen the process of establishing national identities.

Shankar, after his choreographing and performing experience with Anna Pavlova, began creating his own dance by presenting the 'orient' to the west. Soon he was making a journey of self-discovery and experiencing movements essentially belonging to an Indian tradition. His philosophy of choreography was never contained by borders between the east and the west and constantly moved across to incorporate other ideas and images. His film *Kalpana* (meaning literally 'Imagination') showcased many of his ideas, his views on life, and a lot of his choreography and remains a document of his creative ability to transcend borders. Even though his dance still continues to be described as hybrid, Shankar's dance vocabulary of movements was born out of his encounter with India at many levels – from rural to urban, from male to female, from different ethnic groups, from moneyed businessmen to students from lower-middle-class families.

The issues of local versus global, regional diversity, statehood, modern policies and trends of education, artistic freedom, and political as well as activist intervention were dealt with not from within the structure of nationalistic discourse in this movie but from outside, as Uday remained a protagonist who never was seen as 'one of their own' by the nationalists in the process of building modern India (Sarkar Munsi, 2008: 86).

Uday Shankar's western encounter started almost 30 years before his making of *Kalpana*. In fact, to quote his wife, Amala Shankar, 'Uday felt he was already too old to perform while doing *Kalpana* – and he wanted the dances to be danced by a group rather than be the focus himself'.[2] This was in direct contrast to the fact that he had no other male dancer in his troupe until 1935, when he brought Madhavan from the southern part of India, who was trained in Kathakali and mostly did the roles of 'the tribal or warrior'. During the first several years of forming his troupe and even later, he himself did all the roles of the divine, the god, and the principal male dancer himself. His dance experience with Pavlova, and his immense popularity with women, gave him the confidence to posit himself as the representative of that very orient, which in the colonizer's imagination remained largely a feminine one, in many ways justifying the eclipsing of the feminine orient by the patriarchal, strong white colonial power as a natural rule of patriarchy.

The notion of the 'orient' created a homogenized identity for countries with distinct communities and cultures – most of the time grouping together a completely heterogeneous group of people under the category of people who were spiritual, mysterious, exotic, erotic, and ultimately feminine in their servility, which made them all the more enticing. Mysticism and mystery ruled the choices made of representation in the performances of Ruth St. Denis, Maud Allen, Anna Pavlova, and so many others. At that moment in history, the entry of a male dancer – albeit a handsome and attractive one – was sensational. As Amala Shankar often recalls,

> After a performance, and after innumerable curtain calls, when the clapping was going on and on, Uday stepped forward to acknowledge the appreciation, to fold his hands and say, 'thank you'. One woman sitting in the front row exclaimed in awe, 'He talks!'[3]

Uday Shankar toured the west extensively from 1928 to the late 1950s and became known as the cultural ambassador of India, much to the irritation and anger of the classicists within India. Joan Erdman (1997) writes:

> Suddenly there was an exotic oriental dark (but not too dark) dancer, who appealed to women. In the 1930s, while touring Uday Shankar and his company in the United States, Russian impresario Sol Hurok noted that Shankar's audiences were filled with women, who adored him. And, in turn, Uday adored women, who offered themselves to him frequently and openly. Shankar's major patrons were women, not surprisingly.

In the realm of European and American images of the exotic oriental, Shankar's appearance on the Paris dance scene in the 1930s, and his huge success in France and Germany, as well as America, paralleled a fascination with Eastern spirituality and philosophy. The Theosophical Society was gaining followers in India and abroad. During the time that Uday Shankar's father, Shyam Shankar Chaudhury, was a Sanskrit scholar in Benares at the turn of the century, he became a follower of Theosophical Society leader Annie Besant. Uday's main partner in his first company was Simkie, whose mother was a member of the Paris branch of the Theosophical Society. In addition, a number of European women of various descent and experience, though none of them Indian, had promoted themselves to the Paris public as Indian dancers. Probably the most famous – and later notorious – was Mata Hari, who presented herself as a devotee of Shiva at the Musee Guimet in 1905. So when Uday Shankar appeared as an authentic Indian, but an accessible one, able to enter the demi-monde and other Paris society as a Brahmin, son of an Indian princely state's foreign minister and a former partner of prima ballerina Anna Pavlova, this was an entirely different presence, with legitimizing credentials in place. The fact that he found high-class patronage, sponsorship, and venues for his programmes was a function of both his genius and talent as well as his connections. From 1930 to 1942, Uday and his troupe gave 889 performances all over Europe. During this period, he also visited America several times with his troupe.

In 1937, he performed in Kolkata, and his performance was a huge success. Given the anti-colonial, progressive movements of the time, the mix of dance choreographies, made up of everyday issues of human life and grand mythical themes, was received with great fanfare. Rabindranath Tagore himself welcomed him and later, in 1939, also encouraged Uday to establish an institute called the Uday Shankar India Cultural Centre at Almora, in the foothills of the Himalayas. With an impressive list of patrons like Nehru, Gandhi, Tagore, and Elmhurst backing Uday Shankar, he established the academy for excellence in performance arts. He invited the best of the master teachers to work at the centre, like Amobi Singh for Manipuri, Shankaran Nambudri for Kathakali, Kandappa Pillai for Brahatanatyam, and Ustad Alauddin Khan for music. Shankar's brothers, including Ravi Shankar, were all there at the Almora Centre, which soon became not only the meeting place for various forms of Indian dance and music but also the base of Shankar's performing troupe, which toured and performed extensively in India and abroad. After four years of operation, the Centre had to be closed down due to lack of funds.

The Uday Shankar India Cultural Centre trained and produced many great dancers. Zohra Segal was one of the instructors of the centre – and she also framed the curriculum of the teaching programme there. Some of the people who were at the Centre as troupe members, students, and so on are Zohra Segal and her sister Uzra, Guru Dutt, Simkie, Shanti Bardhan, Amala Nandi (who later became Shankar's wife), Narendra Sharma, Ravi Shankar, Rajendra Shankar, Sachin Shankar, Prabhat Ganguly, Mohan Sehgal, Devilal Samar, Sundari Sridharani, Bhagabhan Das, and many more.

Zohra Segal, a part of the teaching faculty and the performing troupe in Almora with Uday Shankar, in an interview with Kapila Vatsyayan, documented by the Indira Gandhi National Centre for the Arts,[4] speaks of the teaching methods used in Almora, the disintegration and closure of the Uday Shankar India Culture Centre because of World War II, and the drying up of funds coming from foreign patrons. Zohra also talks about the growing interest in the medium of films by Shankar, which led him to plan his only film, relocating and situating himself in Madras. Zohra's observation and admiration are important in the context of the whole preparation and training process of an entirely new group of performers by Shankar after his previous group of students, and the performing troupe was completely dismantled in 1944, as they give us insight into the making of the film.

The huge personal popularity and the image of the performing troupe of the Centre, made up of Uday as the principal male dancer (with a group of female and a few male dancers) is the background to the decision to make *Kalpana* the movie. The idea of making a movie had been in Uday's mind for some years. His painful experience of failing to secure funds for the Almora Centre and his frustration at the refusal of patrons to continue to believe in his venture of creating an institution made him write a story of Udayan, an artiste, a dreamer, weaving in his dance creations in a story which reflected, often simplistically, the images of the colonized, shackled country and the dream and aspiration of the soon-to-be-independent nation along with his own idea of art, creativity, freedom, and ideal patronage. The ideal, in most of the film, remains melodramatically projected through idealistic stories, woven mostly with Shankar playing the hero/protagonist in the story, and the god, the lover, and the handsome dancer in the dream/dance sequences. Joan Erdman (1997) writes:

> Nationalism and colonialism are not mere external contexts for Shankar's dance and success; they are the context in the midst of which he performed, was reviewed, met his patrons, and created his repertoire.

She continues (1997) about Shankar's experiments in dance and the critical response of the audience to these endeavours:

> In evaluating and analyzing Shankar's opus, the same complexities of time are involved. What was modern in the 1930s is historical today. Was his legacy to be in touch with his times? In that case, his son Ananda and wife Tanushree Shankar are carrying on his father's new tradition in their Calcutta company. Or was his choreography so far ahead of its time as to be understood only now? Recently, viewing Uday Shankar's *Kalpana* in Madras (now called Chennai), I sat amongst a dance audience from the Sri Krishna Gana Sabha, which presents an annual dance conference every December. An Italian practitioner of *Brahatanatyam* and scholar of Indian culture exclaimed while watching Shankar's dances, 'This is new, it's post-modern, it's Indian! Everyone needs to see this now'.[5]

After more than 60 years of having remained in the grey zone of oblivion, *Kalpana* is in the news again. The great cinema director Martin Scorsese has restored it. The media woke up to the importance of this film as a significant modern intervention as a result of the involvement of Scorsese's name in the issue. *Hindustan Times* (May 5, 2012) reported under the headline 'Martin Scorsese restores Uday Shankar's classic Kalpana',

> On May 17, Kalpana, the only film dancer-choreographer Uday Shankar wrote, directed and acted in, will probably be screened in the Classics Section at the Cannes Film Festival. The movie was restored by Martin Scorsese's World Cinema Foundation.
>
> *(Roshmila Bhattacharya, Mumbai)*[6]

For me, as an anthropologist, and having grown up as a miniscule part of the Shankar's dance legacy, this interest spells the beginning of a resurgence of interest in Shankar's work in terms of its holistic contribution, his seemingly apolitical and often criticized understanding of the medium of art, and most importantly his idea of nation and citizenship.

Having been constantly pained and surprised throughout the 80s and 90s about the almost complete negation of Shankar's contribution through his dance and dance-related works, it is at this point that I realize the importance of viewing Shankar's work academically in the context of the dual discourses of colonial and post-colonial development in performance and trajectories of development of dance in India.

Kalpana has been ignored to a large extent, as was its maker, as he was more or less put in a box by reformists and nationalists, as none of them were comfortable with categorizing him and his work into slots made for all 'Indian' dance categories, that is, folk and classical (Sarkar Munsi, 2010: 31). As a feature film and a chronicle/document of Uday's work, the film, made at a critical juncture of the period of five years from 1944–1948 and released in 1948, speaks of many facts and has never been properly analyzed for its structure, narrative, politics or the lack thereof, or any other issue, as there has been no academic work on this. In order to start the discussion on the movie in general, I would like to discuss the time frame of the making of *Kalpana*.

The film *Kalpana* is one of Shankar's most celebrated/critiqued achievements and an important part of his creative legacy. Released in 1948, *Kalpana* was made at a time which marks the point of transition from the era of the 'Raj' to the formation of the new nation-state – India. Made through the period of handover of power by the colonial power to the new Indian nationalist polity, the film stands as an important narrative amongst others, recording ideals, aspirations, and imaginations for the new nation – at one level, of course, embodying and reflecting the artiste's subjective imagination but at another level reflecting the grand narrative of the predominant Hindu cultural statements made through the imaging of the new nation by its political planners. *Kalpana* forms a cultural bridge between two epochs with its narrative and dance sequences as imagination (which is the meaning of the word 'kalpana'). The film is a testament to Shankar's dual artistic legacy and often-critiqued sensibility, reflecting his interpretation of his cultural heritage combined with an acquired sense of stage presentation due to his western exposure. Erdman writes:

> Shankar went to Madras to make his only film, Kalpana, which he showed in India and abroad in 1948–49. Praised for its dance filming, critics and audiences found Kalpana's story troubling. It was a pre-independence narrative and contained reminders of what was alleged to tarnish Uday's reputation. In brilliantly creative dance scenes and less scintillating dramatic ones, Kalpana satirized Indians who tried to retain lost power after independence and showed women in competition for his favors.
>
> *(1997)*

Often criticized for his borrowed western eye and aesthetics, Shankar, from the beginning of his choreographic journey, had two distinct elements in his choreography and choice of narrative. On the one hand, his projection of his imagined India, coming alive through popular Hindu myths and portrayal of gods and goddesses, with costumes and jewellery carefully crafted and often chosen from different adjacent cultures of Bali, Java, and so on, created an extended 'Indian'-ness, extended beyond the everyday 'real' to the glorified, exoticized, aspired-to Indian life. This element is abundant in *Kalpana*, as Shankar chose to weave the most popular of the dances into the main narrative of the film. Inevitably he danced as Shiva or Kartikeya in all these dances.[7] In the reclaimed and restructured classical dances like Brahatanatyam, gods and goddesses were also main elements. But in Uday's productions, the themes around the gods assumed a grand dimension beyond the straightforward portrayal of the popular mythological stories, with his use of group choreography.

Structuring the image

Shankar worked through his film text by choosing certain issues and developing his dance sequences (or mostly fitting his existing choreography) within them. The narrative worked on two levels in order to create images of the 'existing' and the 'imagined'/'ideal' nation. At the level of existing reality, the great tragedy of the Bengal famine, the transition from a feudal/agrarian to an industrial society with a different faces of oppression of the ruling class, and the emergence of an elite, moneyed class – which had the power as well the voice of authority – carried the main thread of continuity of the storyline. This 'real' was juxtaposed with a constant and at times indistinguishable transference to the dream where the imagined 'ideal', mythical, magical, or the supernatural, was woven in. This was where Shankar incorporated all his popular dance creations around mythical themes, not always caring about sequencing them in a relevant manner with the storyline. Here he also wove in his own idea of the imagined independent nation – addressing issues of the land–man relationship, education and women's emancipation, different phases of male–female relationships, the ideal structure of an institution for teaching art and patronage, and so many other things. I would like to focus on specific scenes to analyze the ideas and images that were central to the making of this film.

The turmoil of transformation from an agrarian to an industrial mode of life

A sizable portion of the film is devoted to the question of industrialization and the right balance between agriculture and the new industrial growth. Starting with an individual entrepreneur showing and talking about a model of a factory that he imagines will change the life of the people completely, this scene takes us to a complete mechanization of human efforts and energy and finally a revolt by the workers against exploitation by the owners of the means of production. The narrative is simple but direct, where the workers lament the complete misuse of land and the neglect of agriculture as a means of livelihood. It also stresses the importance of a balanced approach to agriculture and industry.

The extensive sets used in the scene where Shankar takes up the issues of urbanization, industrialization, the exploitation of the work force, human life as machine, and rural–urban balance are a reworked section from his choreographic work *Labour and Machinery*. Using simple wooden boxes of different dimensions and projections of parts from a typewriter, he planned and created sets of the semi-dark interior of a factory, using mechanical movements to depict the mechanization of life. More noteworthy than the sets or the depiction of the theme is the range of movements – created for the representation of the total scenario of machines and the workers – not drawn from any particular existing movement vocabulary but actually created for the specific purpose of enacting the scene. Extensive use of choreographic designs at different levels and complex synchronizations of multiple rhythms as well as movements establish the activities of a factory and the changed life of the people, who are still in a dilemma over the choice of industrial work over agriculture.

In this particular scene, a striking balance is achieved between movements from the daily lives of ordinary people and postures and techniques from different existing movement vocabularies of Indian dance – thus enhancing the creative moments and also providing a huge range of non-grammatical movement patterns which until then were never thought to be possible ingredients for creating a dance. Movements from daily life, such as eating, shaving, walking, sitting, and running, have been utilized in two particular ways: first to build a stylized movement vocabulary out of everyday movement practices and second to incorporate the everyday movements into the story as tools to help the depiction.

The women in Kalpana*: the image of/for the nation?*

A large part of the critique of Uday Shankar's movie at that time came from within India. His storyline was criticized on the ground that it portrayed Udayan, the principal protagonist, played by Shankar, being surrounded by women. Some harsher views also mentioned Shankar's attraction to women and their overawed reaction to his presence on stage. The source of discomfort could also be about the way the female roles were framed. Women's roles in *Kalpana* are crafted largely on the basis of Shankar's stage productions – portraying the ambiguity of women's position in the minds of the nation-planners (Sarkar, 2001). On the one hand, a woman remains the 'perfect' woman; the mother; the 'ideal' emotional, possessive, yet submissive lover. On the other hand, in several scenes, she emerges as an educated equal partner of her male counterpart, challenging age-old shackles put around females in the name of tradition and also defying discomfort about the female body viewed through the idea of what Susan Manning calls 'the male gaze'(2003) and 'public' spaces.

Throughout *Kalpana*, women appear in specific roles as well as in group performances, as a part of the main narrative, and also in the group choreographies. As mentioned before, the principal female roles are woven around the character of Udayan, as two female characters vying for his attention. This part of the story drew a lot of criticism, as the characters and the story appear weak and in sharp contrast to the independent, self sufficient women who become an important part of the dream sequence.

On a discursive level, the nation's ambiguity about its women is reflected in *Kalpana*, in the way the whole question of gender has been handled. In the context of nationhood, women in India have been largely visualized and projected as the carriers/reproducers of culture and ideologies. Women in the context of the popular nationalist discourse are seen principally as mothers, and therefore their role as the socializers of children has been highlighted time and again. Hence, all responsibility for a nation's upliftment is indirectly or directly the women's. The nationalist discourse has always set the 'ideal' role for its women citizens, thereby creating separate standards of social/cultural behavioural practices for male and female subjects. On yet another level, women are portrayed as the symbolic embodiment of the nation as a gendered entity, the eternal mother, who absorbs all the hurt, disgrace, and calamity directed towards her children, who is pained when the children quarrel amongst themselves and create disharmony. This much-used image becomes a common denominator of women in later Bollywood movies of the 50s and 60s as well and continues to be the framework within which the ideal notion of women's role/citizenship operates in the country.

Dreams of development vs. existent social structure

Shankar also included a scene which had less dance and more dialogue, questioning the means and ends of the existing education system, where a group of male and female students wearing graduation gowns come out in two gender-specific lines. The scene hints at the implied equalizing effect of modern education where the telltale signs of class, caste, gender, religion, and so on are overshadowed by the effect of the formal degree. Shankar's hope of a caste-/class-/genderless society becomes clear here where he puts emphasis on modern education as being the path to development. At the same time, his apprehension about the misuse of an educational system, which could create more inequality if the goals are not clearly defined, is also conveyed. The students are shown to be walking out as their fathers are waiting. The fathers start expressing their delight at their sons' rise in social status as graduates and their increased market value, by saying things like, 'Great, my son, you are so qualified that I can ask for fifty thousand rupees for your dowry now!' The continued scene portrays the young generation's confusion and response at their fathers' reactions and at the end of education by all of the graduates pointing to a huge question mark on the screen behind them and saying to the fathers in unison, 'We respect you, but this is our future'. This is followed by a crowd of female graduates, all with their degrees rolled up in their hands, expressing their concern about the means of education. The use of vernacular languages instead of Hindi marks the dialogue where they shout in a visible act of protest, 'We do not want such education' and 'It is impossible to build a nation without a national education policy'. The scene culminates in all the students throwing their degrees up in the air in a clearly portrayed act of protest. The next continued sequence is of the same group of women, dressed in urban, everyday clothes (with a deliberate projection and assertion of independence in the choice of clothes; the walk, with the head held high, without any cover on their heads; and very obviously self-assured body language), walking down what looks like a fashion show ramp of today. They face a group of older men, whose traditional garb hints at their social position as conservative, who try laying down rights and wrongs to the women. This sequence, known to the dancers as 'old and young', was performed by us as members of the performing troupe of the Uday Shankar India Culture Centre, as a part of the recreated choreography *Labour and Machinery* by Amala Shankar in the late 1970s. It had powerful moments of questioning of age-old norms, which still continue to plague women in the name of tradition. It is a liberating moment where women challenge, question, and ridicule the old men by using dance and facial expressions and also breaking into loud laughter each time the men try to assert the 'dos and don't dos'. While stemming from questions regarding education and social status, this scene adds to the dialectic development of the image of the woman in general in the film, referring again to the ambiguity about the woman's position as felt even in contemporary India.

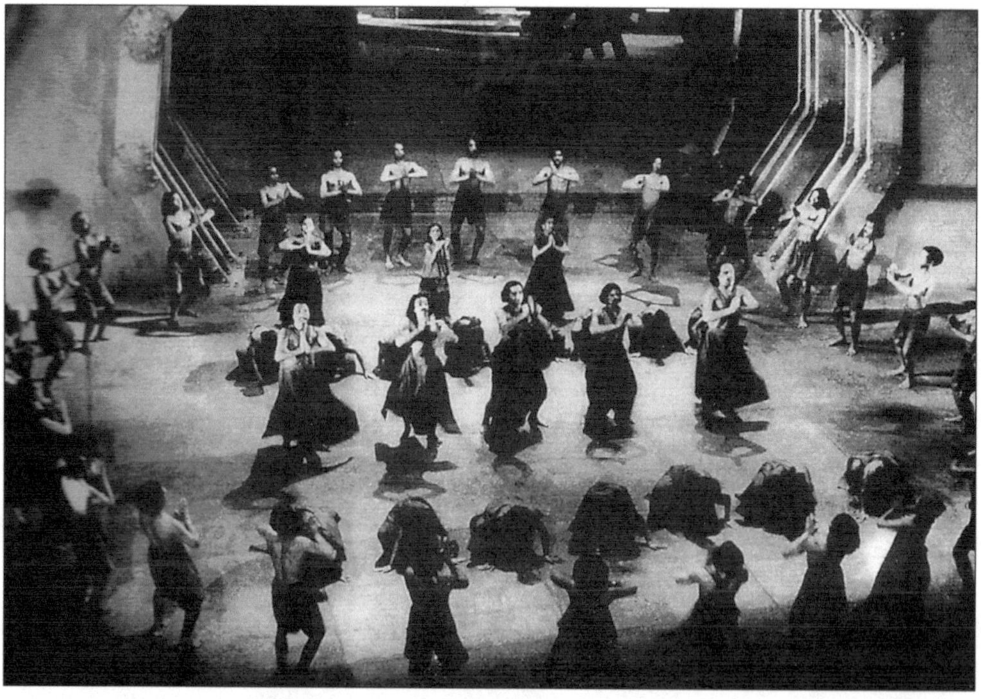

FIGURE 7.1 'Labour and Machine', scene from *Kalpana*

Source: Photo courtesy of Amala Shankar

Exaggerated and stylized movements of the body which are much beyond the natural movements of protest are used along with the everyday natural movements, just as in the rest of film, interspersed with structured dance moments. All three of the previously mentioned elements are carefully choreographed to make special sense of the dialogue not only in the language of films but also in the choreographic language specific to dance.

The much celebrated theme of 'unity in diversity'

The film makes constant reference to the much-hyped theme 'unity and diversity', which became the principal nationalistic projection of the Indian state with its hugely diverse population. The celebration of caste, class, ethnic, and religious diversities takes up a sizable amount of the 122-minute film, where diversity is celebrated through a proud montage of dances by different ethnic and religious groups in their traditional dresses and ornaments, using their own musical equipment and songs. The reference to diversity comes up time and again at different moments and remained a favourite theme of Shankar's dance repertoire before and after *Kalpana*. In fact, his popular repertoire based on dances of India became a popular trend followed in Indian Peoples Theatre Association (IPTA) performances and countless other performances of later years.

Reality vs. dreams: the problem of metaphors

Another particularly important issue that is taken up and woven into the story is the relationship of the state to the private kingdoms of the local princely states. The inclusion/involvement/legislative power concerning these princely states were an important matter debated throughout the last phase of the colonial period. A satirical take on the freedom of these princely territories addressed the autonomy of these states and their heads both at a realistic and an absurd level. This particular issue – woven into the dreams and aspirations of the protagonist of *Kalpana* – given the time frame of making the movie, is particularly political in its choice, as it is not a part of any of Shankar's existing dance choreography. His concern about the ongoing negotiations for integration and the British policies for these states was reflected in the film. Partha Chatterjee, in his introduction to his edited book *State and Politics in India* (1997: 1–2), writes:

> There were some 565 princely states over which the British exercised paramountcy without actually incorporating those territories into the provinces of British India. According to the terms of the transfer of power, the lapse of British paramountcy meant that the rulers of the state regained full sovereignty, although they were given the option of joining either India or Pakistan. There was furious diplomatic activity on the part of the new political authorities of India and Pakistan in the days immediately preceding independence to get prices to sign the instruments of accession to their respective dominions. Vallabhbhai Patel, the deputy Prime Minister of India, took initiative in this regard to put together a single territorial entity over which the newly independent Indian state would exercise sovereignty. The states were first asked to concede to the Indian Union only the powers of defence, external affairs, and communications, and were invited to continue participating in the upper house of the Dominion legislature where a new constitution was being made. In the end

> most of the princes of states surrounded by or contagious to the territory of India – 554 states to be exact – agreed to join.

Let us take, for instance, a scene about Udayan's dream institution in which a huge number of people have arrived for a public performance. The princes of private princely states start arriving on their palanquins and special carriers. Many people, both Indian and foreigners, in western clothes arrive alongside scantily clothed peasants, villagers, and workers. The gates to the auditorium are low. People have to crawl in. When asked the reason for this, one of the volunteers explains the necessity to bow down to art and artistic endeavours – regardless of their social or economic positions. The satire and comedy is made evident in the way the heads of the princely states are dressed or are exercising their power – or through their reactions and implied positions at being subjected to common Indian legislative regulations. Shankar's personal opinion about the special powers of the princely states and the national debate around these issues are incorporated in the imagination through Udayan's dream. During the same performance, Indians wearing western clothes are barred from entering and are asked to come back wearing national costumes, while the westerners wearing the same kind of clothes are allowed in. The role of an ideal citizen is etched out alongside the expectations from the soon-to-be-independent nation.

Shankar's concern with and questioning of the power structure and statehood, along with many other immensely relevant issues in the context of the time the movie was made, of course make the film a document and record of protest, resistance, and debate. But, more importantly, it also highlights the penetrating farsightedness of the storyteller/choreographer/director, which made him focus on issues that emerged as debates of tremendous importance in the following years and continue to trouble the country and its policy-makers.

There are constant dialogues, narratives, and songs which raise questions like 'who does this nation belong to?', 'what is the way in which all people, male/female, old/young, poor/rich in India will have equal rights as citizens?', 'what is the end of education?', along with slogans and statements like 'The land belongs to the ones who toil in it', 'there cannot be a nation without national education', and 'be selfless for the mother-nation' which are projected in the context of dramatized storylines portraying the dream.

Throughout the movie, Shankar acts as the protagonist, trying to provide a simple solution/answer to all the questions coming to the fore in the matter of policy decisions regarding the shape and structure of the soon-to-be-independent nation. Shankar's Brahmin, upper-class upbringing and his extensive exposure to the west did not make him very different from many of the policy-makers, who also came from similar backgrounds. But his stand and critique of the directions that the national policies were taking placed him on the opposite side of the turf from the planners of the nation.

Is *Kalpana* an important part of the dance history of the nation?

Institutionalized categories of dance remain curiously detached from the day-to-day issues and life-changing events in India, continuing to be largely unaffected and unreflective of processes and voices of change that are taken up in theatre, cinema, and visual art. Some dancers, who work on issues beyond the usual ones, feel more comfortable calling their performance dance-theatre. Addressing issues of gender, class, conflict, and resistance directly or indirectly through abstraction has remained a marginal activity in dance even today. The works of Chandralekha (Bharucha, 1997),[8] Manjushree, and Ranjabati Chaki Sircar[9]

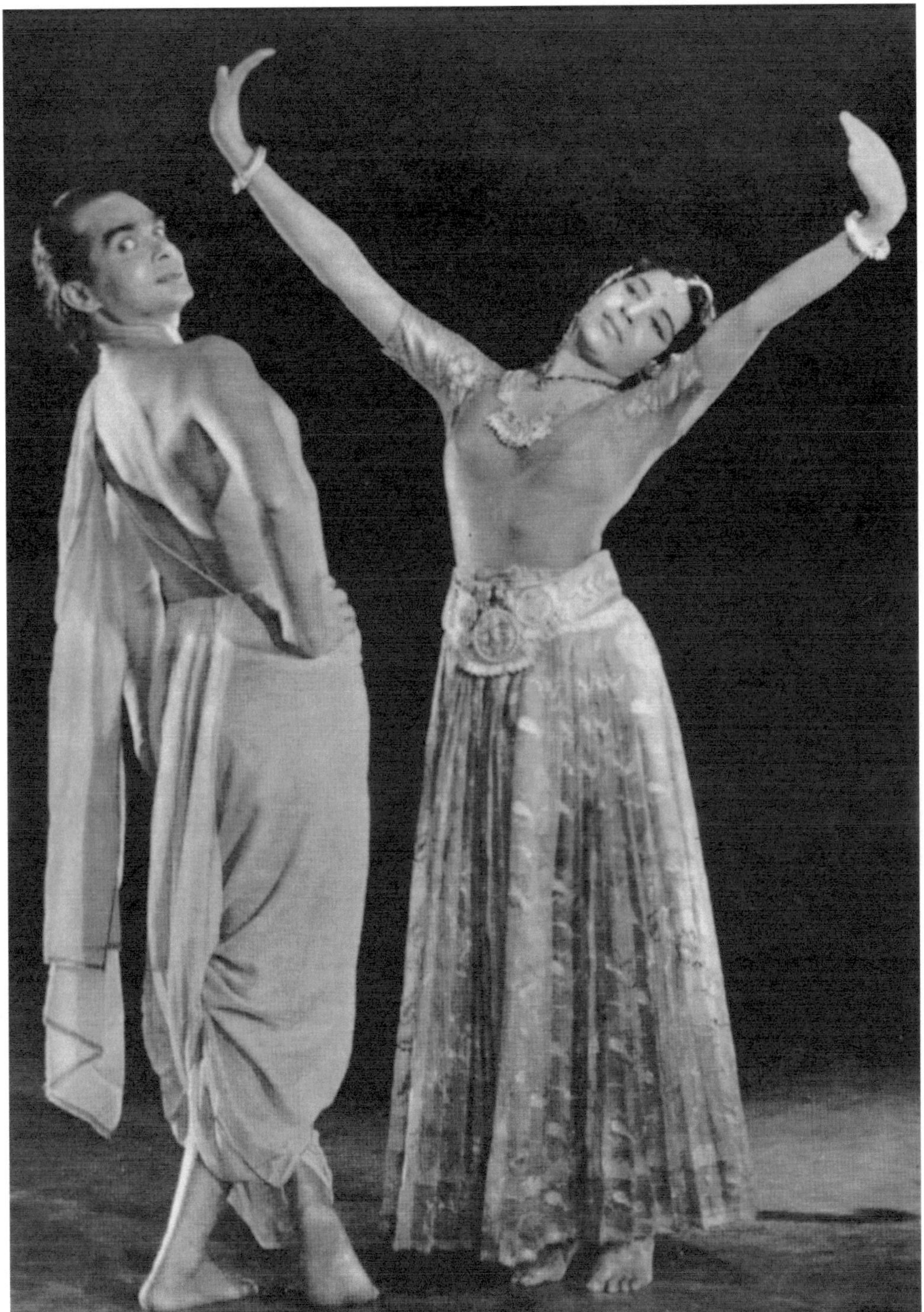

FIGURE 7.2 *Kalpana,* Amala Shankar with the late Sri Uday Shankar

Photo courtesy of Amala Shankar

have been seen and written about as 'unusual' and 'out of the ordinary', as there is uneasiness about the specific socio-political concerns that these choreographers express through their works. What about Shankar's *Kalpana*? Where does one place it in dance history? Or is there even a need to do so? Does *Kalpana* need to be assessed as a milestone – where dance was used as tool for creating a resistant voice within the nation? Or was it a resistant voice at all – and how so? Dance in India remains largely comfortable within the mythic and mystic realms, where, to date, references to *Shiva* as the creator of dance in the Indian context become a part of the pre-performance announcements – while talking about the 'real' history of Indian dance. Traditional mytho-social themes dominate dance choreographies and solo presentations. Considered against the backdrop of such a scenario, *Kalpana* is modern beyond imagination – in the context of its creation, as well as in the context of today.

If one overlooks Shankar's problems with film as a medium, does *Kalpana* actually address and reflect the question and struggles of the voiceless? As the issues of state, nation, strife for self-governance, independent statehood, various new degrees of exploitation, industrial land use and resistance from within communities, Special Economic Zones and the political struggle, instances of army brutality, and protests take a deeper toll on peace and sovereignty issues of the country today, a more engaged reading of *Kalpana* is necessary as the issues taken up in the movie – sometimes fleetingly and sometimes with immense detail – reappear in everyday life of the nation which is now in its 63rd year of independent existence. It is also important to assess Uday Shankar's work as a part of the modernizing processes in Indian dance in line, continuity, and context of all the developments since – and not as a film, in the context of film history in India, with all its problems in the hands of a first-time director who was working solely on his own instinctive knowledge of the medium to give shape to his imagination.

Note: This chapter was originally published in Sarka Munsi, Urmimala and Burridge, Stephanie (ed.) (2010) *Traversing Tradition: Celebrating Dance in India*, Routledge, India.

Notes

1 Kapila Vatsyayan, in her research, chooses to call the Indian classical dances 'neoclassical' (1995: 486), given the new and constructed history that justifies their inclusion as classical.
2 Amala Shankar, who performed the character of Uma, the main female protagonist, and the love interest of Udayan (portrayed by Uday Shankar himself), discussed the making of the film on several occasions with her students.
3 Numerous discussions over a long time between me and Smt Amala Shankar have always brought out the element of wonder at the stage presence and presentational skills of Shankar as a performer, choreographer, and presenter – and the tangible sense of awe of the western audience.
4 *Zohra Unmasked*, Great Masters Series, documented by Indira Gandhi National Centre for the Arts, Part 1 & 2, contains Zohra Segal's discussion with Kapila Vatsyayan on herself as a dancer as well as an actress and also sheds light on the establishment of the Uday Shankar India Culture Centre in Almora and her association with it.
5 Erdman, Joan L., in her paper 'Who Remembers Uday Shankar?' presented at the annual conference of the Society of Dance History Scholars, Barnard College, New York, on 21 June 1997, discussed Uday Shankar's dance works and their acceptance, critically looking at his vision and achievements in the socio-cultural context of the Indian dance scenario of the time.
6 The *Times of India* reports on 4 of February 2010: 'Martin Scorsese, the director of American Classics Mean Streets, Raging Bull, Taxi Driver and Good Fellas on, and one of the founding members of the World Cinema Foundation dedicated to the preservation and Restoration of film classics, has decided to take up the cause of Uday Shankar's dance-ballet on celluloid, Kalpana'. timesofindia.indiatimes.

com/entertainment/TOI Mobile. Accessed 3 May 2010. The renewed interest of the famous director suddenly triggered a lot of reaction in print as well as audiovisual media from film personalities, though the dance world and the government remain silent to date. On 5 of May 2012, *Hindustan Times* carried the news of the completion of the restoration under the headline 'Martin Scorsese restores Uday Shankar's classic Kalpana'.

7 In later years, to date, some of the dances were never recreated, as Amala Shankar felt that it was impossible to imagine anyone else in Uday's place in these dances.

8 Chandralekha's intervention (Bharucha, 1997; Chatterjea, 2004) was at the level of the body as the tool – bringing in her concerns about the sensual and sexual – and working on a parallel aesthetics, tracing elements of understanding from yogic and tantric references, as well as the cosmic elements of body and dance movements. Analysing Chandralekha's dance, Rustom Bharucha (2008: 3–18) writes, 'Chandra's inventory is better served if it can challenge us to engage creatively with her points of reference to body, society, cosmos, and resistance, and to all those in-between states of consciousness and being that she loved to tease out against the grain of fixed categories and dichotomies'.

9 Aishika Chakraborty's book *Ranjabati: The Dancer and Her World* (2008) and Alezandra Lopez Y. Royo's essay 'Classicism, Post-Classicism and Ranjabati Sircar's Work: Redefining the Terms of Indian Contemporary Dance Discourses' (2003) tell us about the genre of modern dance that Ranjabati Sircar, with her mother, Manjushree Chaki Sircar as her principle teacher, worked on developing as a tool for expression and communication rather than a package of learnt and transmitted idioms. She took up issues of war, feminism, the environment in the choreographies created either solely by her or by the mother–daughter duo within the short span of her dance career, brought to an unfortunate end by her death when she was only in her 30s.

References

Bhaba, Homi. (1990). Introduction to Homi K. Bhaba ed. *The Nation and Narration*, pp. 1–7. London: Routledge.

Bharucha, Rustom. (1997). *Chandralekha: Woman, Dance, Resistance*. New Delhi: Harper-Collins.

Bharucha, Rustom. (2008). 'Remembering Chandra: Points of Reference in the Here and Now', in Urmimala Sarkar Munsi ed. *Dance: Transcending Borders*, pp. 3–18. New Delhi: Tulika Books.

Chakraborty, Aishika. (2008). *Ranjabati: The Dancer and Her World*. Kolkata: Thema.

Chatterjea, Ananya. (2004). *Butting Out: Reading Resistive Choreographies through Works by Jawole Willa Jo Zollar and Chandralekha*. Connecticut: Wesleyan University Press.

Chatterjee, Partha. (1993). 'National History and Its Exclusions', in *The Nation and Its Fragments*, pp. 110–115. Princeton: Princeton University Press.

Chatterjee, Partha ed. (1997). *State and Politics in India*. New Delhi: Oxford University Press.

Erdman, Joan L. (1997). 'Who Remembers Uday Shankar?', This paper is a revised version and was originally presented at the Annual Conference of the Society of Dance History Scholars, Barnard College, New York, on June 21, 1997. Originally published in Interchange, a Publication of Dance Alloy. Website: www.mukto-mona.com/new_site/mukto-mona/index.htm. Accessed: 29/03/2009.

Manning, Susan. (2003). 'The Female Dancer and the Male Gaze: Feminist Critiques of the Early Modern Dance', in Jane C. Desmond ed. *Meaning in Motion*, pp. 153–166. Durham, NC, USA: Duke University Press.

Menon, Sadanand. (2005). '"Passport Please!" Border-Crossings in the Invented Homelands of Dance', in Mohd Anis Md Nor and Revathi Murugappan eds. *Global & Local: Dance in Performance*, pp. 27–43. Kuala Lumpur: The Cultural Centre, University of Malaya.

Pathak, Abhijit. (2006). *Modernity, Globalization and Identity: Towards a Reflexive Quest*. New Delhi: Aakar Books.

Renan, Ernst. (1994). 'The Question of Definition', in John Hutchinson and Anthony D. Smith eds. *Nationalism*, pp. 17–21. New York: Oxford University Press.

Royo, Alessandra Lopez y. (2003). 'Classicism, Post-Classicism and Ranjabati Sircar's Work: Redefining the Terms of Indian Contemporary Dance Discourses', *Roehampton Research Papers*, http://roehamp

ton.openrepository.com/roehampton/bitstream/10142/12584/1/royo+classicism.pdf. (accessed 31 October 2007).

Sarkar, Tanika. (2001). *Hindu Wife, Hindu Nation: Community, Religion and Cultural Nationalism*. New Delhi: Permanent Black.

Sarkar Munsi, Urmimala. (2008). 'Boundaries and Beyond: Problems of Nomenclature in Indian Dance History', in Urmimala Sarkar ed. *Dance: Transcending Borders*, pp. 78–98. New Delhi: Tulika Books.

Sarkar Munsi, Urmimala. (2010). 'Another Time, Another Space', in Pallani Chakravorty and Nilanjana Gupta eds. *Dance Matters*, pp. 26–39. New Delhi: Routledge.

Vatsyayan, Kapila. (1995). "The Future of Dance Scholarship in India", in *Dance Chronicle*, Vol. 18, No. 3, pp. 485–490.

Audio-visual material

Zohra Unmasked, Great Masters Series (Kapila Vatsyayan interviews Zohra Sehgal), Documented by– Indira Gandhi National Centre for the Arts, Part 1 & 2.

8

THE BEAUTY MYTH AND BEYOND

Looking at the Bollywood 'item number'

Priyanka Basu

Introduction

The history of dance in Bollywood is intricately linked to other performing art forms like theatre, music, and several indigenous and classical dance forms. Dance genres in Bollywood, both in their folk and classical adaptations onscreen, have undergone constant mutations to allow the genesis of an entirely new genre of dancing, the Bollywood dance. Over the last decade, numerous critical sociological, cultural, and film studies analyses have assessed the evolution and place of Bollywood dance globally, especially pertaining to the pedagogical and performative contexts outside India and South Asia.[1]

The form of dance analyzed in this chapter is the 'item number' that has remained an integral part of the success recipe of Bollywood films. In its ultimate analysis, an 'item number' is no particular dance per se. It does not pertain to any one specific dance form, either Indian or Western, and hence, it can be seen and portrayed as a site of appropriation of a number of dance styles. The item number is a novelistic discourse in the carnivalesque spaces of Bollywood, as opposed to the epic stature of the 'classical' in Indian dance.

In speaking about the 'item number' per se, this chapter considers several issues that become veritable components of a larger package called the 'item number'. The 'ideal' performing body, the problematics of the 'beauty myth', the agencies of female performers (and to a certain extent also the male item dancer) become the parameters for placing the 'item number' in the reference frame of a 'contemporary' form of practice.

Looking back: the Beast and the Frog Prince

Undoubtedly and inevitably, of the many stories that have featured in schoolday story-reading sessions, the stories of *The Beauty and the Beast* as well as *The Frog Prince* have remained necessary favourites. Both of these are fairy tales guided and connected strangely by a single thread, the thread of 'beauty'. For both the Beast and the Frog Prince, it becomes imperative to appease their lady loves in order to shed their 'ugly skin' and emerge as the beau straight from a Mills and Boon romance. They live happily ever after, and yet there

DOI: 10.4324/9781003160007-14

remain fissures and fragments within this discourse of 'happiness' which probably would not have existed if the 'transformation' had not taken place.

This chapter does not aim to take a detour and interpret these stories in light of structuralist/post-structuralist analyses. Instead, it is an endeavour to borrow the key themes of 'beauty' and 'transformation' against the backdrop of these narratives and analyze how the 'item number' is fed by the mutual play of these categories, both historically and sociologically speaking.

To re-look at 'beauty' as a demand and a judgment upon women, one needs to look at *The Beauty Myth: How Images of Beauty Are Used Against Women* by Naomi Wolf (1991). However, Wolf's views are not only restricted to cultural understanding or shaping women as sex objects but also men as success objects. Between female liberation and female beauty, Wolf states:

> The more legal and material hindrances women have broken through, the more strictly and heavily and cruelly images of female beauty have come to weigh upon us. . . . During the past decade, women breached the power structure; meanwhile, eating disorders rose exponentially and cosmetic surgery became the fastest-growing specialty . . . pornography became the main media category, ahead of legitimate films and records combined, and thirty-three thousand American women told researchers that they would rather lose ten to fifteen pounds than achieve any other goal. . . . More women have more money and power and scope and legal recognition than we have ever had before; but in terms of how we feel about ourselves physically, we may actually be worse off than our unliberated grandmothers.
>
> *(Wolf, 1991, p. 10)*

Wolf's argument centres solely around the Western woman who is made to succumb to the norms of patriarchy that have historically transformed her role/image from that of the *feminine mystique* to the *iron maiden*. The beauty myth, a political way of maintaining the patriarchy, allows women to enter the labour force in a controlled way. The beauty myth replaces the public interest in women's virginity with the shape of her body. Women's magazines like *Vogue*, which came into existence in 1969, focused more on the image of the 'perfect' body that made every housewife aspire to it. 'Beauty pornography', as Wolf terms it, showcases underweight models in the age group of 15–20 years. Wolf's statistics reveal the huge number of women who die every year in the United States due to anorexia and bulimia – the 'beauty myth' operates on the other side of aspiration towards the 'perfect' body.

The Beauty Myth provides one with an entry point into the main concern of this chapter – the Bollywood 'item number'. Item girls in the Indian scenario are not much removed from the realities of the beauty myth, nor is the 'aesthetics of aspiration', as Pallabi Chakravorty discusses in 'Remixed Practice: Bollywood Dance and Global Indian' (2010), bereft of any means to 'not harness' the performer's body. The 'size-zero' female and the 'metro-sexual' male exist not very distantly from each other in a closer scrutiny of the body-image-centric politics of Bollywood.

How does the scheme of the item perform? Within the rubric of dancing bodies, the performers become sacrificial victims to a growing cult of the body. The 'beauty myth' works on the idea of the 'perfect body' and the craze for the size-zero or six-pack ideal. Performance in such a packaged form is nothing but comportment highlighting the best that one's body can be. There is a demand for the 'perfect body', and there are means to cater to such demands. To understand the nexus of the 'beauty myth', 'transformation', and the 'perfect body', it is imperative to take into consideration the idea of the female performer and her position in the socio-economic scenario of a space as amorphous as Bollywood. The following section traces the woman performer and the shifts in her roles within the broader shifts of the history of Hindi cinema and Bollywood.

The 'good' and the 'bad' woman: building convenient images

The 'angel' and 'monster' dichotomy that remains the meta-point in seminal works on female sexuality, like *The Madwoman in the Attic* (1979) by Sandra Gilbert and Susan Gubar, finds constant expression in the history of films in Bollywood. Ever since its inception and even after Independence, Bollywood has tackled such issues as caste politics, widow remarriage, nation-building, and so on. On the surface, such films had little or nothing to do with the subject of female sexuality. Deep down, however, the dichotomy of the heroine or the 'good woman' and the vamp or the 'bad woman' existed with all its paraphernalia. The virgin/whore binary upheld the heroine as the pinnacle of traditional Indian virtues, while the vamp was the 'other' – westernized, scheming, career-oriented, and bold. The heroine or the chaste woman found herself featured in songs that formed the expression of love by the hero. She was the unimposing, coy, and docile love interest of the male protagonist. This very role, however, underwent transformations in the 'dream sequences', where the male protagonist could happily imagine the woman in wet saris, giving suggestive glances. Opposed to all this was the vamp or the 'fallen woman'. Through years of such storytelling, Bollywood danced out the good and the bad, or the virgin and the vamp binaries – always assigning certain typifying movement principles for certain sequences, moods, or characters. However, that was before the time of the 'item numbers'.

Within the reference frame of globalization, glocalization, and the shrinking of spaces to form *scapes*, the item number cannot be dealt with without the subject that is veritably the politics of commodification of the body of the woman performer. An indirect descendant of the 'tamasha' in the play-houses, or, in the Marathi popular tongue, the 'pilay' houses, the item number accepts both the popular indigenous as well as the urban popular 'bar dancing'.[2] It is the in body of the woman that the gazes of the male audience of both the rural/semi-rural as well as the urban popular merge and culminate. In a site like this, the item performer and, more specifically, the item girl, is doubly marginalized – with respect to gender biases as well as the question of morality. At a time when there is a hue and cry against the practice of bar dancing based on the labels of 'moral' and 'immoral', the item number ironically becomes a tool that places the real on the reels so as to allow audience gazes to shift to a site which is less slippery. The 'immoral other' is recuperated in the salver of nation-building or the 'national identity' which Bombay cinema has always been a part of.

It is in this process of appropriation of the 'other' that the economics and overseas marketing of the 'item number' becomes an easier process as and according to the frameworks of the 'woman of the nation' and the 'beauty myth'. The woman performers' body is much more saleable and marketable in this process of mirroring or upholding the national identity to the diasporic audience for whom Indian cinema always become a process of identification with the roots. Needless to say, the 'good woman' had already existed behind the evident garb, whereas the 'other' in the 'item girl' is wrapped up in discourses that are political, social, and economic underneath dance movements that have been appropriated both from the 'classical', the 'western', and the 'non-classical' repertoires. Analyzing the item number allows one to show how the body can be the site of social performance, resistance, and aspiration, thus playing on the notions of morality. It also makes us aware of the fact that 'classicization' or the process of classifying is not a one-dimensional teleological structure emphasizing one single theme but voicing variously to contribute to the formation of a national/trans-national/global identity.

This analysis is further informed by concepts such as 'alterity', 'binary opposition', and the body as a social construction (borrowing from disability studies inaugurated by Lennard J. Davis).[3] Such concepts in relation to the analysis of item numbers are entirely post-modern, keeping in mind the amorphous qualities of post-modernism itself. 'Alterity' as a term is defined as the 'otherness' of people, events, and ideas that are radically different from or of those who are in a position of power.

> The idea of the binary opposition is a corollary of alterity because the cleaving of any narrative in to two sides, one in opposition to the other, is the way in which the nature of alterity in that narrative can be detected. It is such a binary, such a detection of alterity that operates in the disability studies of Lennard J. Davis. He suggests that the human bodies have been binarised into normal and abnormal, ideal and less-than-ideal by forces starting from the eighteenth century eugenics to today's Hollywood movies. The disabled body is therefore marginalized, just as the able-bodied is centred.
>
> *(Chatterjee, 2007, p. 206)*

In most films, where female performers like Helen, Bindu, and Aruna Irani were given catchy, upbeat, sensuous songs, preferably within the setting of a club, pub, bar, or restaurant, the dances were popularly known as cabaret rather than being item numbers. The basic difference between a cabaret and an item number lies in the fact that they are not always the forte of the vamp of the film. A cabaret was integrated into the theme and action of the film; it helped the story to move forward – be it 'Piya tu ab to aaja' from *Caravan* (1971), 'Dilbar dil se pyaare' from *Caravan* (1971), or 'Mehbooba Mehbooba' from *Sholay* (1975). An item number does not necessarily have to be an integral part of the plot of the film. An item number defies the authority of anything that can be deemed the plot of the film. It is a separate discourse in itself, sometimes having little or nothing to do with the main story of the film. Placed within the opening or the closing credits of a film, an item number allows the performer a specific space to assert his or her individuality removed from the heavily pompous sets of Western locales or Indian mansions.

The *itemization* of dance

An item number lacks the elegance of the main narrative; it has the propensity to appropriate every other form within its novelistic discourse. An item number violates the classical, and yet it provides a space for the classical to adjust itself within the item number's polymorphous domain. An item number is a space for perfect fusion or blending of forms. It does not restrict itself to any particular movement, form, or gesture, either masculine or feminine, and forms an altogether different dance style. Before item numbers came into being, sometimes film narratives progressed through dances. Many times, the hero or heroine grew up through the course of the song and dance. In the past, there were always some heroes or heroines who could be utilized better in dance sequences than others – such projection (seen as eroticization) of the male or female was a process that had begun long ago. As Ashok Row Kavi says about Shammi Kapoor:

> Shammi Kapoor achieved this eroticization through a highly personalized form of high camp behaviour and acting which made him the focus of attention. The heroine in all of his moves was a silent, still, and slightly startled persona who drew all attention away from herself through a very low-profile performance. Meanwhile, dancing and prancing around her, in a mixture of plastic and erratic movements would be the hero, as he redirects the audience's attention to his highly vibrant, sexualized performance.
>
> *(2000, p. 309)*

With the shifts in audience tastes, economic mappings within a global phenomenon, and transformation of Bollywood as a pervasive phenomenon, the role of the hero underwent major and discerning changes. Just as the role of the female counterpart to the hero was constantly sidelined, it also began to flow with an altogether different tide with the ushering in of the global audience. The creation of a metro-sexual male was based not on the lines of the Renaissance man or the national icon but the idea of the perfect male body. In this endeavour, dance rose to become the prime tool of showcasing the ideal body. One can cite the example of Prabhudeva and the awe-inspiring flexibilities of his moves. Prabhudeva, however, could not accommodate himself within the mainstream of Bollywood despite being a dancer of the first order, a prerequisite of the film actor. To be appropriated into the rubric of the marketable Bollywood, it is essential to appropriate oneself into the beauty myth or the folds of the perfect body. Songs and dances in Bollywood highlight this prerequisite in an exorbitant way, more so through the aid of the item number.

Songs in Hindi films fall under certain broader categories: festival songs, emotive songs, dream sequences, love songs amidst nature, or comic relief songs. A closer scrutiny would reveal that an item number does not adhere to any of the previously mentioned categories, and yet it can assimilate all these forms within itself. Urmila Matondkar's 'Aa Hi Jaiye' and Sonali Bendre's 'Mujhe Sajan ke ghar jana hai' in *Lajja* (2002) exhibit two varieties of festive spaces within the narrative of the film, both being item numbers at the same time. An item number has an entirely different set of paraphernalia where the dominant modes and equipment of mainstream songs are set aside. The item numbers are discourses in themselves, being removed from the main plot of the film and yet retained as an integral part. Analytically, an item number has always been a binary opposite as far as the film is concerned.

In his *The Scenography of Popular Entertainment*, Brooks McNamara says:

> In the entertainment environment, it is the events that are fixed, and not the spectator, who is often completely mobile and presented with a large number of choices around which he organizes his own 'event'. Entertainment environments are of two types: those that redefine an already existing area for a short time and those that make up what might be called a self-contained autonomous environment. . . . Self contained or autonomous environments use spaces solely devoted to entertainment.
>
> *(McNamara, 1974, p. 23)*

It is this self-contained or autonomous environment that upholds a form like the item number. When, on the one hand, we can problematize terms such as 'spectator' so as to permute a fluid category as the 'spect-actor', why not problematize the performer per se as far as the item number is considered? Taking, for example, item numbers like 'Kaal Dhamaal' in the opening credits of the film *Kaal* (2005) or 'Bhoolbhulaiya' in the closing credits of the film *Bhoolbhulaiya* (2008), these are solely devoted as audience-pullers rather than offering any progress or ultimatum to the narrative of the film.

To read into the word 'item', it is used more frequently in classical Indian dance, where the danseuse is said to perform one 'item' after another on stage like *Jatiswaram* or *Krishnashabdam* in *Kuchipudi*, *Tillana* or *Dashavataram* in *Brahatanatyam*, *Pallavi* or *Ashtapadi* in *Odissi*. Hence, the word 'item' pertains to more of a presentation that is short and crisp in nature. In an item number, too, the length and crispness of the performance is paid special attention so that it does not consume so much of the audience attention as to mar the original effect of the film's narrative. However, Bollywood item numbers, in many instances, have remained longer in the popular imagination and memory than evidently the main narrative of the film.

> Dance not only reflects what is but also suggests what might be. In this process dance generally both mirrors and prefigures shifts in sexual mores and gender roles. Changing dance style and dramatic theme are cultural constructions that respond to an altered environment, encompassing the views and behaviour of women.
>
> *(Hanna, 1993, p. 132)*

The shifts in the roles of the heroine in a Bollywood film reflect themselves in the item numbers as well. To be precise, it is the item number that portrays the changing gender roles. Bollywood actresses of high stature and greater repute do not hesitate to perform in single item numbers without bagging a role in the movie. It is thus a way of asserting the female presence in the predominant Bollywood environment of male bonding.

A *parody* in contemporary usage refers to works created to mock, comment on, or poke fun at an original work, its subject, or author by means of humorous or satirical imitation. As Linda Hutcheon (1985) puts it, parody is an imitation with a critical difference, not always at the expense of the parodied text. Any cultural practice which provides a relatively polemical allusive imitation of another cultural production or practice is generally termed a parody. Item numbers are the perfect site for parodying. Taking a look at Shahrukh Khan's item number 'Hum hai rahi pyaar ke' in the movie *Rab ne bana di Jodi* (2009), one finds a parody of the superstars of yesteryear – beginning from Raj Kapoor and passing from Dev Anand to Shammi Kapoor to Rajesh Khanna and ending with Rishi Kapoor. This item number is carefully placed in the middle of the film within the discourse of another film that the heroine goes to watch in a movie theatre. The larger-than-life effects of the movie hall, the epic grandeur of the sets, and the grandness of the hero as well as the heroine are hearkened back to as well as parodied in the course of the song – a process that could be only possible through the agency of the item number. Item numbers thus provide a space for going back to the 'golden era' of Bollywood and borrowing song sequences that have always remained in the popular memory – songs which the audience would like to visit and revisit again and again. Kareena Kapoor's 'Yeh Mera Dil' in the remake of *Don* (2006) or the appropriation of Helen's item number 'Mehbooba Mehbooba' from the film *Sholay*

(1975) in as many as two recent films like *Ram Gopal Ki Aag* (2007) and *Aap Ka Suroor* (2007) reveal the fact that item numbers are not decadent and that they can be performed and incorporated in any given space of a Bollywood film.

Two important instances of item numbers can be cited from the film *Dostana* (2008) – one in the opening credits of the film by Shilpa Shetty entitled 'Shut Up and Bounce' and the other in the closing credits of the film entitled 'Maa da Ladla Bigad Gaya'. *Dostana* is a film with an overtly and entirely queer subject. The initial item number that seems to portray the dancing diva Shilpa Shetty in her beachwear is actually a misnomer that finds real revelation in the final item number. That the film is entirely about an underlying text of male bonding and that the female space is obviously negligible is heightened by the lyrics of the final item number, 'Maa da Ladla Bigad Gaya', proving that there wasn't any female space throughout the film despite the portrayals of Shilpa Shetty and Priyanka Chopra. Interestingly, the final item number is also featured as a full-fledged song in the middle of the film strictly pertaining to the progress of the events. An item number is novelistic in the sense that it can be taken up in any form to produce an altogether different discourse. The song in the middle of the film entitled 'Maa da Ladla Bigad Gaya' is sung from the point of view of a distressed and scandalized mother, although in a comic vein. However, at the end of the film, when the song resurfaces again as an item number featuring Abhishek Bachchan, John Abraham, and Priyanka Chopra with occasional inclusions of all the minor characters from the film, it becomes dialogic, plural, and polymorphous, with all the voices speaking about the queer theme with their multifarious reactions.

The item number also has the potential to include the 'classical'. Vidya Balan's 'Aami Je Tomar' in the film *Bhool Bhulaiya* grammatically conforms to *Kuchipudi* and *Bharatnatyam* (Kothari, 2001), movements and yet has become one of the identifiable markers of the film as an item number. The item number allows the actor or actress within the film to come out of the role he or she plays and perform as an individual performer. The item number thus becomes occasional comic relief or an interlude during the length and breadth of the film. In most cases, however, such interludes are provided by Bollywood females tagged as 'item girls', who do not belong to the dominant mainstream queue of actors and actresses. Malaika Arora Khan's 'Chhaiya Chhaiya' in *Dil Se* (1998), 'Mahi Ve' in *Kaante* (2002), and 'Hoth Rasiley' in *Welcome* (2008); Shilpa Shetty's 'Main Aayi Hoon UP Bihar Lootne' in *Shool* (1999); Shamita Shetty's 'Sharara Sharara' in *Mere Yaar Ki Shaadi Hai* (2002) and 'Chori Pe Chori' in *Saathiya* (2002); and so on feature in the middle of the films' narratives, sufficing as adept interludes. However, item numbers by superstars or former Miss Worlds and Miss Universes are also featured in the prime moments of the film, like Aishwarya Rai's 'Ishq Kameena' in *Shakti* (2001) and 'Kajra Re' in *Bunty Aur Babli* (2005). For actresses of the second rung like Isha Kopikkar, Rakhi Sawant, Koena Mitra, Yana Gupta, Amrita Arora, Lara Dutta, Negar Khan, and others, the item number is the only available route to a name and fame. The item number is all encompassing, providing space to superstars and budding stars alike. Item numbers become handy for actresses like Esha Deol and Amisha Patel who have not managed so far to feature in any blockbuster film except one or two. For such actresses, the item number is definitely a means to survive rather than resurrect. For models-turned-actresses like Bipasha Basu, item numbers like 'Beedi' and 'Namak' in the film *Omkara* (2006) become an integral part of the role in the film. Item numbers, thus, may be used as accessories or an undetachable part of the main discourse of the film.

From dancing bodies to 'bodies'

What is the element of dance in Bollywood? How does dance cater to the expectations of the audience? One is reminded of the exquisite sets and lavish props of *Devadas*, particularly in the song-dance sequences. Some of these were most redundant to the plot of the film, fitted into the adapted narrative for the sake of a saleable, marketable showcasing of dancing bodies. Item numbers followed from the exquisiteness of the lavish dance sequences – a phenomenon that would extend from the swaying Meena Kumari in *Pakeezah* to the lilting Rekha in *Umrao Jaan* or Madhuri Dixit in *Devdas*. With changes in the spatio-temporal realities, dance burgeoned as a tool for the performer. The male gaze became as true as the female gaze and vice versa.

In considering the item number or the item performer, it is important to consider dance both as a movement and a metaphor in the 'item' per se. Bollywood is the idea of 'perfection'. This is not in line with the 'classical' view of perfection but a sense of perfection centring solely on the look of the body. There are methods of creating magic moments – editing; extra fast/unreal speed; rhythmic, exercise-like structure; and dances structured out of beautiful bodies and looks and some not-so-important techniques – out of a rote repertoire, sometimes borrowed, or even more seldom created. These are the photo-finish moments rather than movements occasionally incorporating certain Indian dance moments. In so doing, Bollywood progresses with its own movement logic – keeping in mind the realm of the possible – given the lack or short experiences of training of the item dancers. Dance movements depend on the expertise of the dancers, but the choosing of the dancer is not dependent on how she dances but what her screen presence adds to the film in terms of sexuality, sensuality, and so forth. Digressing into a regional (Bangla) dance competition (aired on a fresh Bengali channel, *Rupashi Bangla*) for the Bengali housewife, one is able to witness the metamorphosis that a *Boudimoni* (colloquial for a Bengali middle-class wife) undergoes in order to emerge as the dancing diva showcased on television. Season 1 of the programme ran successfully under the name of *Tolly vs. Bolly* (Tollywood, the film industry of Kolkata; Bolly, a shortening of Bollywood), setting a reference frame for the *Boudimonis* of Season 2 in terms of the perfect body by sheer virtue of which the Bollywood girls/performers in the show outdid the Tollywood counterparts.

The nature and content of dance training and trainer become intrinsic to the process of 'creation' of the item number. Analyzing a dance movement taken from a cross-section of the item number highlights the derivations from 'classical' to movements that are part of a specific genre that is solely Bollywood. Considering fully what Saroj Khan terms the 'signature movement' in Bollywood dance, one can take into account the star-studded item song from *Om Shanti Om*, a conglomeration of various 'signature movements' pulled out and accommodated within the rubric of the song through the agency of the actor/actress excelling in them or those that are entirely theirs in celebration of the success of that particular 'signature movement'. The role of the dance 'guru' becomes inevitable in the context of the Bollywood methodologies of dance. The dance masters of Bollywood need to be understood in a comparative fashion against the dance gurus of classical traditions. Calling themselves dance directors and appearing in all popular TV shows as judges, the Bollywood dance gurus are much more seen and heard than the traditional dance masters, teachers, or gurus; undoubtedly, Farah Khan or Saroj Khan, Shiamak Davar or Vaibhavi Merchant remain as much in the limelight as designers of the 'perfection' of the actor's body as much as the director himself/herself. They also become the powerful, famous, visible persona beyond India and among the diaspora as well, with the economic factor and the flow of money channelizing the soaring of the Bollywood dance masters.

However separated and differentiated the mainstream and the popular are in terms of dance in India, Bollywood has travelled beyond borders and is popularly sought after as a technique to be learnt and performed – so much so that the most popular form of Indian dance in terms of people wanting to learn is Bollywood, and the 'Item' dancers have achieved for the dance something to be admired by people mainly in the age group of 5–25. The regime of perfection, beauty, body, camera effect, and marketability all become the potential to draw teenagers from India as well as outside India, that is, the diaspora. The item number involves largely a Western clientele, the role of which is inherent to the idea, creation, and presentation of the genre. There can be little doubt how and why Indian dance is known in the larger global context as Bollywood dance.

Conclusion

There is a conscious effort in this concluding section to exhibit some of the advertising strategies of the dance maestros of current Bollywood, taking a cue from the previous section on Bollywood dance:

> Shiamak's Indo Jazz. Hip Hop. *Bharatnatyam.* Krump. Kathak. Belly Dancing. Salsa. Broadway. Shiamak's Bollywood Jazz, *Odissi,* Afro Jazz, Shiamak's Yoga, Rock N Roll, Contemporary.
>
> Shiamak's brand new *Winter Funk*™ Show 2009 is a platform for students to perform dances which represent various regions of our diverse planet. Mainstream international dance styles will vie for place with brilliantly choreographed Indian classical, semi-classical and folk arts, along with street styles, tribal and folk dances from across the world. *Winter Funk* this year promises to create a landmark in the dance and entertainment world.
>
> Shiamak ensures that there's something for every dance aficionado on the globe, by contrasting the popular new street style called Krump with a Belly Dancer's sensual undulations, the contemporary dancer's organic moves with a Kathak dancer's complex footwork, and a Salsera's fiery gyrations with earthy Indian folk arts.
>
> Shiamak often says, 'We are all equal, but we are not the same.' Celebrating unity in diversity and the special skill of Shiamak's instructors in presenting such versatility, Shiamak Davar's Institute for the Performing Arts (SDIPA) presents this never-before-seen and much anticipated spectacle.
>
> Shiamak believes that dance has the ability to not only unite your mind, body and soul, but also people across any physical, psychological or cultural boundary. 'Expressing yourself through dance is the purest form of communication. Join us at *Winter Funk*™ 2009 to experience dance like never before,' signs off Shiamak.
>
> This mammoth festival of dance will celebrate over 20 dance styles, besides showcasing Shiamak's Indo Jazz, Shiamak's Bollywood Jazz and Shiamak's Yoga.
>
> Source: https://www.shiamak.com/the-dance-academy.html

B O L L Y W O O D! Known as Mumbai Film Industry, Bollywood is located in the heart of India-Bombay. It is the largest film industry in the world, producing almost double the number of movies and selling a billion more tickets each year than Hollywood.

Along with its glamorous touch, Bollywood films are mostly known for their elaborate musical dance numbers. Producing more than 2500 songs per year, they have also been influential in Hollywood movies such as *Moulin Rouge, Chicago* and *Vanity Fair*. The Bollywood dance style incorporates the fusion of Hip Hop, Jazz, Salsa, East Indian Classical, Folk, and Bhangra dancing.

For the first time in Los Angeles, the true essence of Bollywood Dancing is brought to the United States, by YOGEN. He is the ONLY choreographer in the nation who has worked in Bollywood. Bollywood Step Dance is dedicated to the art of Bollywood Dancing.

Bollywood Step Dance was launched in 2005, and in a short period of time has generated a reputable image in creating a vibe for the Bollywood Style of Dance.

Catering to all the ethnic backgrounds and variety of age groups, Bollywood Step Dance provides professional dance training and troupe performances for stage shows, television, film, corporate events, concerts, wedding reception, private parties, workshops, and social/public events at a national and international level. Our costumes are exclusively designed in Mumbai, India.

Get ready to step into the world of glitz and glamour of Bollywood dancing with ***Bollywood Step Dance!***

Source: https://www.hollywoodbowl.com/musicdb/artists/726/bollywood-step-dance-with-yogen

The most popular of the dance schools in Mumbai promote the teaching of the 'Bollywood' style of dance. The Bollywood style generates the item number and the item performer. Such performers are no longer selected, male and female, from the lower rungs of the social strata struggling for their everyday bread and butter but girls and boys from rich families whose skins glow without mustard oil and whose bodies are shaped in air-conditioned gyms. The manufacturing of 'perfect bodies' in the dance schools of name and fame have sidelined the older Cine Dancers Association. For the marginalized performer or the 'extra', an item number might emerge as a mode of prosperity provided the 'perfect body' is manufactured.

Item numbers become the perfect instrument of crowd-pulling not only for the Indian audience but also for the overseas audience. Such has been the popularity of the item numbers that Urmila Matondkar's 'Chhamma Chhamma' from the film *China Gate* (1998) was even featured in the Hollywood film *Moulin Rouge* starring Nicole Kidman. Whereas Bollywood movies are said to be heavily inspired from the storylines of the Hollywood blockbusters, instances such as these reveal the popularity of pomp and show in the West as far as Indian films are concerned. The Indian view of life is primarily festive. In his *Critique of an Everyday Life*, Henri Lefebvre says: 'Festivals contrasted violently with everyday life, but they were not separate from it. They were like everyday life, but more intense' (1992, p. 207). Item numbers point at the festive spirit of the everyday – the everyday audience, to be specific, for whom reel life becomes real life after the day's ordeal and the everyday performer for whom the 'beauty myth' and its appendage of the 'perfect body' become the necessary evil.

Note: This chapter was originally published in Sarkar Munsi, Urmimala and Burridge, Stephanie, (ed.) (2010) *Traversing Tradition: Celebrating Dance in India*, Routledge, India.

Notes

1 See Bhattacharya Mehta and Pandharipande (2010), Deshpande (2009), and Kavoori (2008).
2 See Morcom (2013) for a discussion on the history and practice of bar dancers.
3 See Davis (1995).

References

Bhattacharya Mehta, Rini, and Pandharipande, Rajeshwari. 2010. *Bollywood and Globalization: Indian Popular Cinema, Nation, and Diaspora.* London: Anthem Press.

Chakravorty, Pallabi. 2010. "Remixed Practice: Bollywood Dance and the Global Indian", in Pallabi Chakravorty and Nilanjana Gupta (eds.), *Dance Matters: Performing India.* New Delhi: Routledge, pp. 169–181.

Chatterjee, Niladri Ranjan. 2007. " 'The Great Little Man': The Body as a Structural Motif in Mulk Raj Anand's Untouchable", in Nandini Bhattacharya (ed.), *Mulk Raj Anand: Untouchable.* New Delhi: Dorling Kindersley (India) Pvt. Ltd., pp. 205–214.

Davis, Lennard J. 1995. *Enforcing Normalcy: Disability, Deafness, and the Body.* London: Verso.

Deshpande, Anirudh. 2009. *Class, Power and Consciousness in Indian Cinema and Television.* New Delhi: Primus.

Gilbert, Sandra, and Gubar, Susan. 1979. *The Madwoman in the Attic: The Woman Writer and the Nineteenth-Century Literary Imagination.* New Haven: Yale University Press.

Hanna, Judith Lynne.1993. "Classical Indian Dance and Women's Status", in Helen Thomas (ed.), *Dance, Gender and Culture.* London: Macmillan, pp. 119–127.

Hutcheon, Linda. 1985. *A Theory of Parody: The Teachings of Twentieth-Century Art Forms.* New York: Methuen.

Kavi Row, Ashok. 2000. "The Changing Image of the Hero in Hindi Films", in *Journal of Homosexuality*, Vol. 39, No. 37, pp. 307–312.

Kavoori, Anandam P. 2008. *Global Bollywood.* New York: NYU Press.

Kothari, Sunil. 2001. *Kuchipudi: Indian Classical Dance Art.* New Delhi: Abhinav Publication.

Lefebvre, Henri. 1992. *Critique of Everyday Life* (trans. John Moore). London: Verso.

McNamara, Brooks. 1974. "The Scenography of Popular Entertainment", in *The Drama Review*, Vol. 18, No. 1, pp. 16–24.

Morcom, Anna. 2013. *Illicit Worlds of Indian Dance: Cultures of Exclusion.* London: Hurst Publishers.

Wolf, Naomi. 1991. *The Beauty Myth: How Images of Beauty Are Used Against Women.* New York: William Morrow.

Films cited

Aap Ka Suroor – The Real Love Story, (Dir.) Prashant Chadha, Benetone Films, 220 mins, 2007.
Bhool Bhulaiya, (Dir.) Priyadarshan, Super Cassettes Industries Limited (T-Series), 220 mins, 2007.
Bunty Aur babli, (Dir.) Shaad Ali, Yash Raj Films, 220 mins, 2005.
Caravan, (Dir.) Nasir Hussain, Nasir Hussain Films, 225 mins, 1971.
China Gate, (Dir.) Raj Kumar Santoshi, Santoshi Productions, 210 mins, 1998.
Dil se, (Dir.) Mani Ratnam, India Talkies and Madras Talkies, 180 mins, 1998.

Don, (Dir.) Farhan Akhtar, Excel Entertainment, 220 mins, 2006.
Dostana, (Dir.) Tarun Mansukhani, Dharma Productions, 225 mins, 2008.
Kaal, (Dir.) Soham Shah, Dharma Productions and Red Chillies Entertainment, 225 mins, 2005.
Kaante, (Dir.) Sanjay Gupta, Film Club, 180 mins, 2002.
Mere Yaar ki Shaadi Hai, (Dir.) Sanjay Gadhvi, Yash Raj Films, 220 mins, 2002.
Moulin Rouge, (Dir.) Baz Luhrmann, Bazmark Films, 175 mins, 1998.
Omkara, (Dir.) Vishal Bharadwaj, Big Screen Entertainment, 190 mins, 2006.
Ram Gopal Verma ki Aag, (Dir.) Ram Gopal Verma, Adlabs Films, 225 mins, 2007.
Saathiya, (Dir.) Shaad Ali, Yash Raj Films, 180 mins, 2002.
Sholay, (Dir.) Ramesh Sippy, Sippy Films, 240 mins, 1975.
Shool, (Dir.) E. Nivas, Shemaroo Video Pvt. Ltd., 180 mins, 1999.

Web pages cited

<https://www.hollywoodbowl.com/musicdb/artists/726/bollywood-step-dance-with-yogen> as accessed May 18, 2021.
<https://www.shiamak.com/the-dance-academy.html> as accessed May 18, 2021.

Indonesia

9

REVEALING CULTURAL REPRESENTATION IN INDONESIAN CONTEMPORARY DANCE

Michael H.B. Raditya

Introduction

As an archipelago country, Indonesia consequently has diversity in its society and culture. According to the Central Bureau of Statistics, it has 1,331[1] tribal groups spread over 17,491 islands: the survey documented the number of tribes, sub-tribes, and sub-subtribes spread across all the territories of the country. This cultural and performative diversity, which has particularly impacted the performing arts, comes from those tribal groups and is always based on the context where the art is made – this reflects a statement from Janet Wolff, who considers art a social product. She states that 'here can be no doubt that any human act is determined. Not only that, but it is multiply determined by social factors, psychological factors, neurological and chemical factors' (Wolff 1981, p. 20). Similarly, art historian James R. Brandon, who was engaged in Southeast Asia, noted that the performing arts in Indonesia had a ritual function. The following statement reflects an earlier era when the society's belief system in the pre-historical era was animism.

> what we can imagine is the existence of a firmly established tradition of folk performances, closely tied to communal rites of animistic worship and propitiation and to cyclic festivals which served the dual purpose of religious worship and entertainment.
> *(Brandon 1974, p. 12)*

Indonesian art historian R.M. Soedarsono agreed with Brandon's statement about ritual uses in society. Furthermore, he identified three functions of art in society: as ritual means, personal entertainment means, and aesthetic presentation (2002, p. 118). In addition to the function of art in society, art historians have also discussed its origin.

Brandon, Claire Holt, and R.M. Soedarsono believed that there were external influences that affected the performing arts in Indonesia, among others, the influence from India from the 1st to 16th centuries, the influence from China and Islam from 1250–1750, and the Western influence from the 16th century to Indonesian independence day in 1945 (Holt 1967; Brandon 1974; Soedarsono 2002). These impacted the trajectory of the performing

DOI: 10.4324/9781003160007-16

arts in Indonesia, encompassing multiple negotiations in its development. Moreover, I agree with Bauer and Bauer that 'not all traditions survive, and sometimes its societal changes cause traditions to vanish' (2007, p. 197). Therefore, some traditions become extinct; others survived, adapted, and created new ones. This chapter articulates the representation of culture in the contemporary dance landscape in Indonesia, investigating each epoch to understand how choreographers engage with culture to respond to current issues.

Tracing the face of culture in the development of dance in Indonesia

Before independence, the performing arts were patronized by the palace, landlords, merchants, or colonialists. After this, the arts came under state power, although some palaces and private patrons still maintained their support. A point to note is that there is a dichotomy in a sociological sense. R.M. Soedarsono (2002, p. 83) surmised that towards the end of the 1940s, the performing arts were attempting to eliminate the boundary between high art – known as classical art – and folk art. The difference is based on the reason behind patronage and ownership. Research indicates the term 'folk art' refers to shows emerging outside of the palace, especially in rural areas; however, the implications of this dichotomy influenced the aesthetic view across both forms of the arts.

The dissemination of power and interpretation via performing arts started to appear, particularly through labeling based on locality, such as Banyumasan style and Semarangan style, which opened up local identities. This agenda of cultural identity was also paralleled in a statement from President Sukarno on Indonesian Independence Day in 1959 that took a stance against Western influences. Sukarno assumed that the massive spread of Western culture in Indonesia had decreased morale and the cultural order of Indonesia. It had an impact on the president's encouragement to focus on local culture. Yampolsky stated:

> the government attempted to establish the Melayu Dance *Serampang duabelas* (in a North Sumatran version) as a national dance that could replace Western couple dancing as a social activity for young people. Another example of Sukarno-era effort to create an instant national culture out of regional materials is the support given by RRI [Radio of the Republic of Indonesia], the state-run radio network, to the genre for music then known as *hiburan daerah* (regional entertainment [music]).
>
> *(1995, p. 706)*

Concerning Yampolsky's statement, I agree with Kusumastuti's point that 'national identity' became the principal framework of artistic development in 1959 (Kusumastuti 2017, p. 777). However, in reality, there was a lack of new artworks that were locally based that did not have an affiliation with the palace. In short, the performing arts and new artworks – including dance – were created for the state's interest.

The performing arts in the post-independence era were also affected by political development in Indonesia. They were utilized in several matters: first, as cultural diplomacy for the government. To present a visualization of the presence of Indonesia in the eyes of the world and to create an official image of national identity, Sukarno regularly sent dancers, musicians, and puppet-masters abroad as artistic ambassadors from Indonesia (Kusumastuti 2017, p. 777). Second, the performing arts were used to attract mass attention by a political party in 1950–1960. In those years, popular theatre forms Ketoprak and Ludruk in Java became a fight between Partai Nasional Indonesia (PNI) (Indonesia Nationality Party) and Partai Komunis Indonesia (PKI) (Indonesia Communist Party) (Soedarsono 2002, p. 93). Interestingly, the art used by the political parties was folk art. Regarding that fact, it could not be denied that in its continuity, these two interests consolidated the contestation between high art and folk art. It often made high art further removed from society, although various efforts continued to show high art in the public space.

Along with this development, in the 1950s, formal education in the performing arts, namely Konservatori Karawitan Indonesia (KOKAR) (Indonesian Karawitan – Musical – Conservatory), began in some places, such as Surakarta, Denpasar, and Bandung. KOKAR as a conservatory not only focused on gamelan but also dance and *wayang* (puppetry). A similar institution was founded in the dance field in 1960, Konservatori Tari Indonesia (KONRI) (Indonesian Dance Conservatory).

The government inaugurated some performing arts tertiary schools, such as Akademi Seni Tari Indonesia (ASTI) (Dance Academy) in Yogyakarta; Akademi Karawitan Indonesia (ASKI) (Karawitan Academy) in Surakarta; Akademi Seni Karawitan Indonesia in Padangpanjang, Sumatera Barat (1965);[2] and Akademi Seni Tari Indonesia in Denpasar (1967),[3] even though the Indonesian economy was relatively unstable. Furthermore, these schools concentrated on Indonesian performing arts in addition to classes offered by the palace and patrons (Maecenas). Besides the educational institutions established by the government, many groups opened dance studios (*sanggar*) in Yogyakarta, Surakarta, West Java, Bali, and others.

Meanwhile, in the new order, the prohibition of Western culture directed by Sukarno was abolished by incoming President Suharto in 1967. The effect was that the influence of Western culture returned to Indonesia. It gradually pushed globalization that affected the performing arts (Soedarsono 2002, p. 100), especially dance. The agenda for performing arts such as cultural missions also continued – it was also utilized as a product for tourism and television (Yampolsky 1995, p. 715). In the Suharto era, Kusumastuti classified dance in 1966–1987 as government support for artists' aspirations (2017, p. 778). The government also brought some creative spaces for the artists. One of them was Taman Ismail Marzuki – this used to be known as the place where contemporary dance started in Indonesia.

Trajectory of Indonesian contemporary dance

The emergence of Indonesian contemporary dance evolved through a long process. It was different from the journey of contemporary dance in the West that included modern and post-modern phases. In contrast, contemporary dance in Indonesia was initiated with a spirit of cultural renewal working from existing traditional dance forms (Murgiyanto 2016, p. 142), and various strategies and negotiations of contemporary dance started.

After independence, Indonesia was connected directly to Western modern dance – especially the American pioneer Martha Graham. This connection continued over several years. It began in 1952 with Seti-Arti Kailola, who studied modern dance for 11 months with Martha Graham in New York (Soedarsono 2002, p. 242). Three years later, Martha Graham staged a public performance in Indonesia. In 1957, Seti-Arti Kailola, Bagong Kussudiardja, and Wisnu Wardhana were invited to study and observe dance at the Martha Graham School and Summer School of Dance. Returning from America, Bagong and Wisnu opened up dance schools in Yogyakarta: Bagong with Kussudiardja Dance Training Center and Wisnu Wardhana with Contemporary Dance School Wisnu Wardhana. Kailola also began a dance school, but it closed in 1964. Bagong and Wisnu creatively negotiated the concept of modern dance they learned in America with Yogyakarta traditional dance. However, Bagong has created an individual dance style that integrates the traditional and the modern (Murgiyanto 1991, p. 151). That is the difference between Bagong and Wisnu. Both schools were the ones that disseminated modern dance to the public. Referring to the term 'contemporary dance', Wisnu is the choreographer who has been using the term since 1958 explicitly, although the idea of contemporary dance did not differ from the modern dance at that time.

The Contemporary Dance School Wisnu Wardahana eventually closed; however, the spirit of contemporary dance was constantly shared by Bagong, and the term 'contemporary dance' was well known ten years later. It was often defined as experimental dance based on the idea of renewal that was different from the previous dance. This idea was established in 1968 at the Jakarta Art Center, Taman Ismail Marzuki (TIM). Progressively, TIM evolved as a space for contemporary dance creation by various choreographers either from Jakarta or other cities. Jakarta became a melting pot not only for culture but also for the formation of Indonesian contemporary dance. The city offered more opportunities and access than other regions; however, it did not mean that other cities were not part of these initiatives.

1968–1990: Reinterpretation of traditional dance: connecting with the present

In the early phase, renewal and working with an experimental spirit were the main inspiration for Indonesian contemporary dance. Taman Ismail Marzuki became a progressive space in the revitalization of dance art in Indonesia and other traditional dance preservation. At Taman Ismail Marzuki, Sardone W. Kusumo, Huriah Adam, Farida Oetoyo, Julianti Parani, I Wayan Diya, Wiwiek Sipala, and others experimented with new ideas. Notably, during this focus on renewal, much of choreography reflected the artists' interaction with traditional dance that became their habitus.

When Taman Ismail Marzuki was formed, Sardono held a workshop followed by Huriah Adam, Farida Feisol, Yulianti Parani, Sentot Sudiharto, and I Wayan Diya. Instead of a workshop teaching the Western modern dance technique that he had learned – for example, from American dancer Jean Edman – Sardono wanted to delve deeper into the technique 'owned' by the choreographers. It was crucial to the foundation of Indonesian contemporary dance, where their own embodied culture became the primary material in creating their works at the Jakarta Art Center, Taman Ismail Marzuki. Sardono W. Kusumo, a choreographer who had experience with cross-cultural dance (starting from Javanese dance to modern dance), was productive in making works in a contemporary way. An essential work of Sardono's in his creative journey was *Samgita Pancasona* (1969). Significantly, Sardono kept on being inspired by Javanese versions of the *Ramayana* – particularly the reliefs on the Prambanan temple – although he presented it as an experimental dance (Murgiyanto 2016, p. 85). In *Meta Ekologi* (1979), Sardono narrated the destruction of the environment (Widaryanto 2015, p. 118). This work became a portrait of ecological consternation in Indonesia. It proved that Sardono's explorations dealt with culture and the contexts that happened around it.

In addition to Sardono, female choreographer Huriah Adam's work was also thought-provoking. Huriah was trained in the techniques of Melayu dance, Pencak silat (a martial art), and Minangkabau dance. Huriah created a work entitled *Sepasang Api Jatuh Cinta* (1968) that was inspired by the 'Tari Piring' – a traditional Minangkabau dance – and martial arts. The change from tradition to contemporary was accomplished by the exploration of dynamic and expressive movement (Kusumastuti 2017, p. 779). Huriah's exploration with contemporary dance did not stop, and she is still an active choreographer; for example, she created *Malin Kundang* in 1969. Huriah reinterpreted a Minangkabau folktale about an ungodly child and his mother. In *Malin Kundang*, Huriah elicited the idea that Malin Kundang's punisher was not his mother but God (Minarti 2020, p. 275). Her belief in maternal morality made Huriah unconvinced that a mother could bear to condemn her own child (Kusumastuti 2017, p. 779).

Farida Oetoyo had a background in ballet. Her choreography *Putih-Putih* (1976) was inspired by the call to prayer and incorporated praying movement (Kusumastuti 2017, p. 781). Although this work was forbidden to be performed because of misunderstanding from certain religious groups, it was notable for linking ballet vocabulary to daily activity in Indonesia. At that time, not only Sardono, Huriah, and Farida but also other choreographers, such as Julianti Parani, I Wayan Diya, Wiwiek Sipala, Agus Tasman, Suprapto Suryodarmo, Dedy Lutan, Tom Ibnur, Gusmiati Suid, and other young choreographers, enlivened the constellation of contemporary dance makers at the time. The presence of senior choreographers still enlivened the contemporary dance constellation, such as Bagong Kussudiardja, who created *Kurusetra* (1987), which linked traditional dance and contemporary spirit.

From some of the previous examples, it is clear that Indonesian contemporary dance was formed from the spirit of renewal, traditional dance reinterpretation, and linking it thematically and aesthetically with the present. Therefore, contemporary dance in Indonesia is linked to culture and current contextual issues. Dance works in this early era kept being produced because of various factors, such as cultural exchanges, workshops, and festivals – both in Indonesia, such as Festival Penata Tari Muda (Young Choreographers Festival), and foreign festivals, such as the American Dance Festival (Murgiyanto 2016, p. 142). These platforms became choreographers' focus to keep on creating works. Festival Penata Tari Muda (Young Choreographers Festival), for instance, became the meeting point for choreographers for disseminating contemporary dance throughout Indonesia.

1990–2010: Reflecting surroundings, expanding the contemporary

The next generation of younger choreographers grew alongside the previous generation of Indonesian contemporary dance practitioners. They learned contemporary dance from performances, festivals, workshops, and discussions that they observed or attended. This second group transitioned from the expiration of Festival Penata Tari Muda – Young Choreographers Festival (1985) to the period of the Indonesian Dance Festival (1992). As they had a direct relationship with the previous choreographers, the goal of continuing what had been established in the initial foundations of Indonesian contemporary dance was maintained; this was re-reading the culture and linking it with the present. Compared to the first generation, this second group comprised a more significant number of choreographers and was spread more outside of Jakarta such that the culture they absorbed as their source material for contemporary dance was more varied and diverse.

The constellation of contemporary dance in this phase still included the previous choreographers, such as Sardono W. Kusumo and Suprapto Suryodarmo, to the youngest at the time, Boi G. Sakti, and the next generation, such as Hartati, Eko Supriyanto, Martinus Miroto, Mugiyono Kasido, Jecko Siompo, Ery Mefri, Bimo Wiwohatmo, Besar Widodo, and others. It could not be denied that this generation also learned contemporary dance from the previous generation and dance schools, either through formal tertiary dance training or from informal dance schools and studios. In brief, they did not start their understanding of contemporary dance in Indonesia from the beginning, but they continued and developed it.

In the second era, I began by introducing one of the female choreographers who consistently created contemporary dance. Hartati – growing up in Solok Selatan, Sumatera Barat – knew Minang dance since she was little. She had martial arts movement and *Randai* (Supriyanto 2018, p. 159) that might enable many possibilities with other dance elements, for example, the combination of ballet and martial arts in work *Wajah* (Muliati et al. 2017, p. 69). Hartati's initial momentum was in *Sayap yang Patah* (2001), telling of her empirical experiences as a woman caught in the middle of domestic issues and a career, confronting these problems in both the private and public domain.[4]

Another choreographer who lives in Sukoharjo, Central Java, Mugiyono Kasido, previously practiced classical Javanese dance. In his journey in contemporary dance, Mugiyono brought the narratives of Indonesian culture, with modern references, into his works (Supriyanto 2018, p. 151); for instance, in *Kabar Kabur*, Mugiyono's inspiration drew from the phenomenon of the fall of Suharto's regime. In the choreography, Mugiyono did a movement exploration with his T-shirt, such as dancing while putting his feet into the sleeve, dancing while hiding his head in a shirt, and so on. The choreography implied that there was always interpretation and resistance within the power (Raditya 2017a).

Another choreographer who has a Javanese classical dance background is Martinus Miroto. Miroto's significant work in his journey was *Penumbra* (1995), his final project performance piece at the University of California Los Angeles (UCLA). Miroto put on a *Panji* mask and danced alone to Javanese classical dance and static music (Supriyanto 2018, p. 144). The Javanese classical dance that was clearly affiliated with the palace met static music – like Jathilan[5] – connecting it with folk art. The combination strengthened the idea of binary oppositions existing in all human beings, such as good and bad, light and dark.

Another significant choreographer is Eko Supriyanto. Eko learned many dance forms from local dances such as *Kuda Lumping* and *Kubro Siswo*, Javanese classical dance, and modern dance. One of the works that came after Eko pursued a master program at the University of California Los Angeles was a contemporary dance entitled *Perang Buta* (*Blind War*). This choreography showed the technique and physical attributes of the dancer rather than relying on a narrative story or drama (Supriyanto 2018, p. 180). Eko combined the movement motifs of *Gagahan Cakil* with classical dance Surakarta style and a release technique from America.

Another choreographer who concentrated on the creation of an innovative movement vocabulary was Jecko Siompo. A choreographer from Papua, he created a movement method known as *animal pop*. It was a combination of primitivism and modernization (Supriyanto 2018, p. 171); for example, in work *Di Depan Papua* and others. In addition to these five choreographers, many others, such as Ery Mefri, incorporated martial arts and Minangkabau dance and Bimo Wiwohatmo, who utilized Javanese dance and daily movement.

In this second era, the development of contemporary dance had innovations in terms of the dance vocabulary, content, and contexts. This era was highly creative and critically thought-provoking compared to the previous era. Open access, and the productive sharing of cross-cultural exchanges, encouraged more innovation and variety. Consistently, dance festivals, both domestic and abroad, became both presentation opportunities and spaces for expression. Dance critic Sal Murgiyanto noted that Indonesian choreographers mostly conducted collaborations and performances abroad – these choreographers included Mugiyono, Sulistyo S. Tirtokusumo, Sardono W. Kusumo, Boi G. Sakti, Miroto, Indra Utama, Ery Mefri, and Eko Supriyanto, to name a few.

On the other hand, dance festivals in Indonesia were more widespread geographically; for example, Contemporary Dance Festival managed by Tom Ibnur in Padangpanjang (1995); Dance Festival for welcoming the 21st century managed by Ery Mefri in Padang (1996); Pasar Tari Kontemporer (Contemporary Dance Market) in Pekanbaru, Riau; and others (Murgiyanto 2017, p. 145). These platforms offered the opportunity for young choreographers to be seen alongside that by the previous generation. Indirectly, they developed the Indonesian contemporary dance scene as a more diverse ecosystem.

2010–2020: Reaching new opportunity and sharpening discourse

In the third era, incorporating cultural dances and experiences as the basic foundation of Indonesian contemporary dance did not change. However, new forms of expression emerged for senior and emerging choreographers in movement innovation, more profound research, and creative exploration. In this case, finding new opportunities in expanding cultural areas that had not been seen before as a dance influence and idea became central. This approach featured young choreographers and the previous generation, for example, Eko Supriyanto, Melati Suryodarmo, Bimo Wiwohatmo, Hartati Mugiyono Kasido, Jecko Siompo, and others.

Eko Supriyanto created the trilogy *Jailolo* that was influenced by his exploration in Northern Halmahera, Maluku, and others: *Cry Jailolo, Balabala,* and *Salt.* In these three works, Eko began with the local culture and linked this with issues occurring in society. In *Cry Jailolo* (2013), Eko processed the movement of *Legu Salai* as a means to discuss the destruction of corals in Jailolo (Raditya 2017b); in *Balabala* (2016), the vocabulary of *Cakalele* and *Soya-soya* was incorporated to present issues for women in Jailolo (Raditya 2018); *Salt* (2018) linked the embodied dance languages from Eko, starting from Javanese classical dance to his experiences in Jailolo.[6] Similarly, Rianto also drew on the dance of *Lengger Lanang* from Banyumas as his dance medium. In *Medium* (2016), Rianto reflected on his movement journey and the dance experiences he obtained through meeting with other cultures.[7] Meanwhile, in *Hijra*[8] (2018 – work in progress), Rianto discusses the problem of gender migration in the cross-gender dance *Lengger Lanang.* In discourse, local and cross-gender issues were not highlighted in the previous eras.

These developments also consisted of the opening up of conversations that were insightful and deep. In this respect, the cultural material incorporated by the choreographers was not new, but critically, they offered specific points of view in responding to a phenomenon, for example, Otniel Tasman, a young choreographer from Banyumas who also had an experience of the *Lengger Lanang* dance that considered masculine, feminine, and the in-between. In *NoSheHeorIt* (2017), gender discourse and performativity became the main idea to be presented.

In addition to Otniel, there are other choreographers, such as Darlane Litaay, a choreographer from Papua. He created *Morning Valley Dark Valley* in 2017. His piece deals with exoticism, centralization, and ecological issues. In his work, Darlane uses several things, such as Papuan dance movements, language, and fashion to criticize the dominance that exploits Papua. Ali Sukri, a choreographer from West Sumatera, created his work *Tonggak Raso* (2016), which talks about 'cultural defense' and protection from external cultures. Martial arts influence his choreography. Ayu Permata Sari, a choreographer from Lampung, was influenced by *Kuadai* dance as movement vocabulary in work *Li Tu Tu* (2018). In *Li Tu Tu*, the discourse was not only about the message of the performance but also the

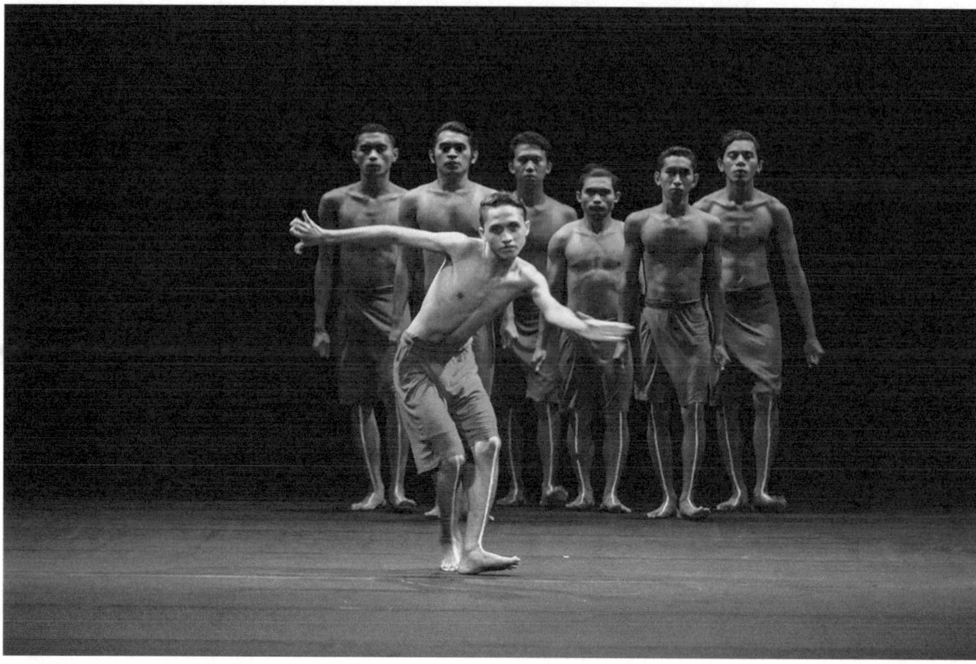

FIGURE 9.1 *Cry Jailolo*

Source: Ekos Dance Company

Eko Supriyanto

Photographer: Feri A. Latief

form of the performance. In *Li Tu Tu*, Ayo showed three elements: critical thinking, exploration, and participatory aesthetics.

In addition to the previously mentioned artists, there are other important choreographers from that era such as Danang Pamungkas, Windarti, Boby AS, Ajeng Soelaeman, Ari Ersandi, Moh. Haryanto, I Putu Bagus Bang Sada, Kinanti Sekar, and others. In the current era of open access, both emerging and senior artists form the cohort of Indonesian choreographers.

Another significant aspect has been the impact of technological advances and openness. This thing has enabled information about festivals, performances abroad, commissioning opportunities, and collaborations, as well as residencies for choreographers, to be easily accessible. Some Indonesian choreographers also became dancers in companies abroad; for example, Danang Pamungkas was a dancer at Cloud Gate Dance Theatre; Rianto performed with Akram Khan in *Until the Lions*; a collaborative dance work between Katia Angel and Ari Ersandi, and more.

These factors all contributed to the aesthetic expansion and contextual depth of the present Indonesian contemporary dance scene. Therefore, the openness of information and the nature of information distribution that was once only one way could now be shared by anyone. Criss-crossing within the country, combined with globalized influences, became the pattern for this epoch. Regarding this pattern, this is the time that Indonesian choreographers explored their habitus and traditions. However, they continually need to beware of the 'cultural exoticism' trap when creating new choreography.

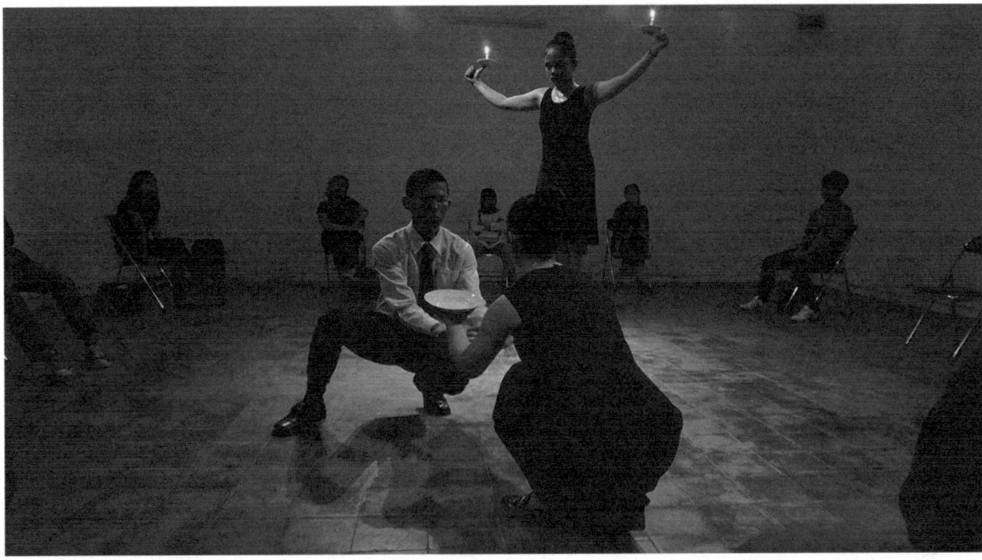

FIGURE 9.2 *Li Tu Tu*

Source: Ayu Permata Dance Company

Ayu Permata Sari

Photographer: Theodore Pulung

Conclusion

The trajectory of Indonesian contemporary dance has undergone rapid development. Traditional arts from various regions became the central material that influenced and inspired cross-generation choreographers. Creators across generations used their culture to express themselves, incorporating themes and ideas ranging from contemporary personal issues to themes inviting wider public reflection. The expansion of cultural influences in contemporary dance was also more intertwined and connected, emerging from the dominant culture to aspects that were not highlighted before. Critical discourse also developed over time, with various perspectives enriching the ideas of Indonesian contemporary dance for choreographers, the audiences, critics, writers, and scholars.

However, all that did not happen individually; cross-cultural meetings and experiences also stimulated Indonesian contemporary dance to be more linked to international discourse and global issues. As a result, Indonesian contemporary dance, with its unique cultural underpinnings, became the medium of continuity, connecting traditional dance to the present and offering contemporary discourse to the world of dance.

Notes

1 www.bps.go.id/news/2015/11/18/127/mengulik-data-suku-di-indonesia.html (accessed 1/07.2020)
2 www.isi-padangpanjang.ac.id/sejarah-isi/ (accessed 1/07.2020)
3 www.isi-dps.ac.id/tentang-isi-denpasar/sejarah-isi/ (accessed 1/07.2020)

4 http://kelola.or.id/seniman/hartati/ (accessed 1/07.2020)
5 Jathilan or horse dances are well known on the island of Java. They are mostly performed in East and Central Java and places where Javanese migrants live today. Horse dances are generally referred to as kuda képang, jaran képang, or kuda lumping (literally 'woven horses'). Jathilan is mainly performed in rural areas. These performances are based on a spirit possession practice known as kasar. Since the dances are not performed as an alus court dance in the Sultan's Palace (keraton), it is not easy to determine how popular the style is (Christensen 2014, p. 93).
6 https://artsequator.com/salt-eko-supriyanto-en/ (accessed 1/07.2020)
7 http://indonesiandancefestival.id/acara/medium-rianto-indonesia/ (accessed 1/07.2020)
8 https://artsequator.com/hijrah-rianto-en/ (accessed 1/07.2020)

References

Bauer, Laurie and Bauer, Winifred. (2007). "Playing with Tradition", *Journal of Folklore Research*, 44 (2/3): 185–203.

Brandon, James R. (1974). *Theatre in Southeast Asia*. Cambridge, MA: Harvard University Press.

Christensen, Paul. (2014). "Modernity and Spirit Possession in Java Horse Dance and Its Contested Magic", in Volker Gottowik (ed.) *Dynamic of Religion in Southeast Asia: Magic and Modernity*. Amsterdam: Amsterdam University Press.

Holt, Claire. (1967). *Art in Indonesia: Continuities and Change*. New York: Cornell University Press.

Kusumastuti, R.A. Siti Nurchaerani. (2017). "The Development of Choreography in Indonesia: A Study of Contemporary Dance Work in Jakarta Art Centre Taman Ismail Marzuki 1968–1987", *Wacana*, 18 (3): 772–790.

Minarti, Helly. (2020). "When She Steps on an Ant, It Will Not Die: But if She Stumbles over a Rice Pestle, It Will Break into Three Parts': An Un-Subtle Subversive Woman Dancing Body in Hoerijah Adam's Dance/Choreography", *Religion*, 50 (2): 260–277. DOI: 10.1080/0048721X.2020.1713517

Muliati, Roza, Udasmoro, Wening and Murgiyanto, Sal. (2017). "Tubuh Yang Mencipta Momen: Praktik Negosiasi Tubuh Dalam Tari Wajah Karya Hartati", *Jurnal Kajian Seni*, 4 (1): 64–78.

Murgiyanto, Sal. (1991). *Moving between Unity and Diversity: Four Indonesian Choreoprahers*. (Doctoral dissertation). New York: New York University.

———. (2016). *Pertunjukan Budaya dan Akal Sehat*. Jakarta: Fakultas Seni Pertunjukan IKJ and Komunitas Senrepita.

———. (2017). *Kritik Pertunjukan dan Pengalaman Keindahan Edisi Baru*. Yogyakarta: Program Studi Pengkajian Seni Pertunjukan dan Seni Rupa, Sekolah Pascasarjana UGM and Komunitas Senrepita.

Raditya, Michael H.B. (2017a). "Menimbang Tradisi pada Tari Kontemporer", *Tribun*, August, p. 15.

———. (2017b). "Tari Cry Jailolo: Bermula Dari Gerak Berakhir dengan Pesan", *Jurnal Ranah*, 5 (1): 45–60.

———. (2018). *Merangkai Ingatan Mencipta Peristiwa: Sejumlah Tulisan Seni Pertunjukan*. Yogyakarta: Komunitas Senrepita dan Lintang Pustaka Utama.

Soedarsono, R.M. (2002). *Seni Pertunjukan Indonesia di Era Globalisasi*. Yogyakarta: UGM Press.

Supriyanto, Eko. (2018). *Ikat Kait Impulsif Sarira: Gagasan yang Mewujud Era 1990–2010*. Yogyakarta: Penerbit Garudhawaca.

Widaryanto, F.X. (2015). *Ekokritiksme Sardono W. Kusumo: Gagasan, Proses Kreatif, dan Teks-teks Ciptaannya*. Jakarta: Pasca IKJ.

Wolff, Janet. (1981). *The Social Production of Art*. New York: St. Martin's Press.

Yampolsky, Philip. (1995). "Forces for Change in the Regional Performing Arts of Indonesia", *Bijdragen tot de Taal-, Land-en Volkenkunde: Performing Arts in Southeast Asia*, Deel 151, 4de Afl. (1995): 700–725.

Malaysia

10

ECLECTICISM AND SYNCRETIC TRADITIONS

Conceiving Malay social dance in the 20th century

Mohd Anis Md Nor

At the midpoint of maritime traffic from the Indian Ocean to the South China Sea, the Malay Peninsula continues to serve as a crucible for cross-cultural activities. Cultural influence from the Indian subcontinent from 300 BC to the 15th century, extensive trade between the Arabian Peninsula and China before the 10th century, the spread of Islam from the 12th century, the voyage of Admiral Zheng He to Southeast Asia in the 12th century, and the advent of European colonial powers in the 14th century have long left their mark on the language and culture of peoples in the Malay peninsula. Throughout a motley of cross-cultural and intercultural practices, Malay social dance, for example, continues to evolve as hybrids, conceived through autochthonous (native to the place where it is found) and/or allochthonous (originating in a place other than where it is found) persuasions and stimuluses.

The formation of the nation-state of Malaysia, comprising the Malay peninsula and parts of Borneo, during the second half of the second millennium has shaped Malay social dance as a multi-cultural crucible of indigenous-Indian-Arab-Chinese-Eurasian hybrids. Over time, one of the most significant contributions of the east-west passage of people and their cultures in the Straits of Malacca and the South China Sea was the geo-political remaking of indigenous performative[1] traditions through the processes of adapting and borrowing alien features into hybridized traditions. Malay social dance is one such example of conceived traditions, invented through the elaborate fusion of multiple influences and experiences.

With its own identifiers that are perceived differently from the European context, Malay social dance evolved through autochthonous and allochthonous cultural experiences, generating an eclectic mix of invented structured movement systems that are often syncretic in form and style within its indigenous performative matrix. Within the performative structure, Malay social dance repertoires have evolved diachronically and synchronically, subtly imbued by loaning and borrowing performative encounters between and amongst the indigenous and foreign populace traversing the region. Through selective processes, Malay social dance repertoires were creatively shaped into specific genres, appropriating allochthonous traditions into autochthonous practices, indigenizing social dances by the local communities. The fusion of eclecticism through syncretic processes in Malay social dance is partly attributed to a conscious process of adapting new ideas into existing dance styles to

DOI: 10.4324/9781003160007-18

enhance and popularize these dances to contemporary taste and expectations. Thus, Malay social dance repertories are receptive to newer movement improvisations and dancing styles whilst keeping to the form and structure that govern each of the genres.

Malay social dances

Dance as a culturally structured movement system may be understood to mean many things in the Malay language. The most commonly used word to describe dance in the Malay language is *tari*. It is not an old term. The word *tari* is use to characterize various forms of arranged or choreographed movement motifs in the ever-expanding vocabulary of modern Malay to denote dance or dancing during the 20th century. In the days before the advent of modern terminology, there were other words for 'dance', for example, *tandak, igal, sayau, totor, gencok, joget, main*, and so on (Nor, 2007, p. 357). All these words emphasize the overarching importance of play and performance. The terms are local and reflect specific forms or styles of structured movement systems peculiar to a region, dialect group, or community. Hence, a structured movement system in a Malay community may be executed in specific forms and styles as long as they fit with the aesthetics and social-cultural milieu of the Malay psyche (Nor, 2009, p. 167). Hence, formally arranged movements as in choreographed pieces belong to the realm of *tari* or *tarian*, which literally means dance performed by trained dancers to be viewed. On the other hand, it could also be regarded as game (*main*) or gambol (*gencok*). Participatory call-and-response singing and dancing are referred to as *tandak*. Malay social dance can either be in the form of *tarian* that is meant to be viewed or as social events to be participated in by all that are present in the form of *main, gencok*, or *tandak* (Nor & Burridge, 2011, pp. 41–42). The former is exclusive and restrictive; the latter is inclusive and embracing.

In spite of the exclusive-inclusive dichotomy, Malay social dance could be grouped into two different categories, which are differentiated by the accompanying musical styles, dance structures, and performance elements. The first category is *zapin*, a derivative of *zaffin*, Hadhramaut Arab[2] social dance, created through syncretic borrowing and adapting to the eclectic performative nuances of the Hadhrami *zaffin*. The second category is the Malay *ronggeng*, embracing three dance genres, the *asli, inang*, and *joget*, performed in sequential and cyclical format. These two groups of Malay social dances are performed in variable styles in different regions in Malaysia. *Zapin* dance is found in almost all of the Malaysian states and is known as *zapin melayu* (Johor), *zapin johor* (Johor), *zapin arab* (Johor), *zapin pekan* (Pahang), *zapin trengganu* (Trengganu), *zapin sindang* (Sarawak), *zapin sebat* (Sarawak), *zapin tidung* (Sabah), *zapin perak* (Perak), and *zapin salor* (Kelantan). The repertories of *ronggeng* may consist of several variants of *asli, inang*, and *joget*. The variations of the *asli* genre are in the form of *senandung, gunung sayang, melayu ssli*, and *dondang sayang* dances. Similarly, the *inang* genre may also be known as *mak inang lama, inang pulau kampar*, and *lenggang mak inang*. The *joget*, on the other hand, may be identified as *tandak, joget lambak*, and *joget moden*.

Syncretizing *zapin*

One of the oldest syncretic forms of folk and social dance traditions in the Malay world is *zapin*, allochthonous in origin, that evolved as an autochthonous tradition. The Malay *zapin* is a hybrid of Arab *zaffin* dance and the music of the Hadhramaut Arabs from Wadi

Hadhramaut in central and southeast Yemen. The earlier migration of Hadhrami Arabs to Malaya, Singapore, and Indonesia from the 15th century to the large-scale Hadhrami migration in the early 19th century to the Indian Ocean and beyond, including Southeast Asia (Ho, 2006), brought *zaffin*, a music and dance form that is exclusively performed and practiced by Hadhrami Arabs and *takhmis* and *qasida* (religious music), a metric composition of a sung poem in Arabic poetry, philosophizing, and religious matters to the Malay Peninsula and Singapore and throughout the islands of Indonesia (Nor & Burridge, 2011, p. 53). The Arab *zaffin* is an exclusively male performance tradition normally performed in Arab quarters, which signifies a patriarchal tradition. On the other hand, the Malays created their own pseudo-Arabic expressions through music and dance after the Hadrahmi Arab *zaffin* of Hadhramaut, infused with Malay dance movements and soundscapes.

Malay *zapin*, which is now known by various other names such as *jipin, jepin, japin, zafin*, and *dana* in Malaysia, Indonesia, Southern Thailand, Brunei, and Singapore, is performed to celebrate events associated with weddings, circumcisions, and Islamic holidays such as *maulidur rasul* (the Prophet's birthday), *eid al-fitr* (the end of Ramadan, the Muslim month of fasting), *eid al-adha* (the Festival of Sacrifice), and *hol* (a memorial fest for departed loved ones). It had taken roots amongst the Malay-Islamic communities in the Straits of Malacca to become one of the most widespread Malay-Islamic folk dance and music traditions in insular Southeast Asia. Like many other Malay folk music and dance traditions, *zapin* became indigenized through royal patronage, engendering new geo-political hegemony and affirming indigenous supremacy over a hybrid invention. Hence, Malay *zapin* is recognized by its regional specificities that are often based on its affiliation with former Malay rulers or Sultanates. In Sumatra, variants of Malay *zapin* are associated with the former Sultanate of Langkat, Deli, Serdang, and Asahan-Batubara in North Sumatra; the Sultanate of Pelalawan and Siak Sri Indrapura in present-day Riau province; the Sultanate of Jambi and the Sultanate of Palembang, which shared historical affinities with the 7th- to 13th-century Sriwijaya kingdom; the Sultanate of Johor-Riau-Lingga in Penyengat, Bintan; and the principalities of Karimun, Bengkalis, and Tembelan (Nor, 2000, pp. 17–61, 249–291).

Zapin music, which usually accompanies the *zapin* dance, is played in three different sections: the *taksim*,[3] a improvised solo played by a single *'ud* or *gambus* (lute) player; the melodic section with *kopak*, a loud rhythmic *marwas* drumming patterns in interlocking style; and *wainab* or *tahtim*, which forms the coda for a piece to end that utilizes an extension of the main melodic phrase and the loud *kopak* drumming pattern (Nor, 2004, pp. 128–130). In spite of strong Arabic influence, the divisional units or sections in *zapin* music closely follow a generic indigenous performative patterns of temporal units marked by drum, gong, or cymbal beats in a musical sound that provides kinaesthetic impulses for a dance or structured movement system.

The musical sections of *zapin* music correspond with the sections of the dance performance. *Zapin* performers are required to enter the dance area in single file or in double rows and present a salutation to the musical prelude or *taksim*, played by a single *'ud* or *gambus* (lute) player. This is to be followed by the linear formation of *zapin* performers, who dance facing one another while repeating dance motifs and tracing a recurring forward and backward floor plan, interrupted with a series of skips and squatting positions, which is also known as the *kopak*. At the end of each performance, the dancers perform jumping and squatting dance motifs to the accompaniment of relatively faster drumbeats in the form of the *wainab*.

Zapin has not only become a music and dance form of the Malays in Malaysia, it has spread far and wide throughout the Malay Archipelago over the last millennium. Today, *zapin* has become a highly respected dance and music tradition among the Malays, who consider it Arab derived and Islamic yet upholding Malay decency, propriety, ownership, and the performative hegemony of the Malay world, or *alam melayu* (Nor, 1993, pp. 29–31). Although the foundation of *zapin* is in the Hadhrami Arab *zaffin*, marked by robust dancing by the men, the syncretic Malay *zapin* is more subdued and relatively more refined. *Zapin* dance and music in the Straits of Malacca became signifiers of eclectic Malay-Islamic performative nuances indigenous to the *alam melayu* of the Straits of Malacca. Shaped through an allochthonous-autochthonous hybrid, the resultant performance tradition represents an embodiment of indigenous creative exploits conceived through the merging two different worlds, the patriarchal Hadhrami Arab-Islamic performative traditions and Malay cultural influences. In spite of the continual presence of two different styles of *zapin* in Malaysia, *zapin arab* (Hadhrami Arab *zaffin*) and *zapin melayu* (Malay *zapin*), the two worlds come together through the processes of engaging Islamic aesthetics while affirming the Islamic worldview and belief system in the Malay maritime communities of the Straits of Malacca. *Zapin arab* is an exclusive dance tradition of the Hadhrami Arab descent groups, while *zapin melayu* is performed by Malays in the Straits of Malacca.

Eclectic *ronggeng: asli, inang,* and *joget*

Another important Malay social dance shaped through an allochthonous-autochthonous hybrid is *ronggeng*, which has an eclectic mix of Malay-European-Asian influence, conventionally performed in cycles of three dance genres, namely *asli*, *inang*, and *joget*. The three dance genres in *ronggeng* dance repertoires are accompanied by a *ronggeng* music ensemble, consisting of one violinist, an accordionist, two *rebana* drummers, and an optional knobbed gong player, playing to the tunes of *asli*, *inang*, and *joget* in successive cycles.

The first cyclic repertoire of *ronggeng* is the *asli* dance. It could also be referred to as *senandung* or *gunung sayang*, and it began from the styles of singing and exchanging *pantuns* amongst village folks, which are referred to as *melayu asli*. An *asli* song is identified by an eight-beat phrase in 4/4 time where the first four-beat phrase has a fixed pattern, while the second four beats are usually improvised but is end-accented by a gong beat on the fourth and eight beat. When danced, the *asli* becomes the most refined of all the Malay folk dances due to the curling and flexing of fingers on the fourth and eight beats while the dancers dance in a slow, walking motion. The *asli* dance was usually performed in the *bangsawan* to accompany singers who sang *asli* tunes. In Sumatra, a musical form similar to the *asli* genre is known as *gunung sayang*. In Malaysia, however, a variation of the *gunung sayang* is found in the musical traditions of the Malay and *Baba* communities in Malacca, Penang, and Singapore, where it is known as *dondang sayang*, or love song (Nor, 1993, p. 40) Today, both the *inang* and *asli* dances have not only incorporated modern and traditional musical instruments but have also included many eclectic musical arrangements played to their respective tunes.

The second repertoire of *ronggeng* is the *inang* dance. Etymologically, the word *inang*, which literally means *wet nurse*, appears as *tarian mak inang*, or 'the dance of mother wet nurse', in the collective *ronggeng* repertoires. The word *mak* is a truncation and also a familiar diminutive form of the word *emak*, which means *mother*. The term *mak inang* refers to the lady-like qualities of the women dancers, whose graceful movements are likened to those of a wet nurse. The dance consists of walking motions either in relatively slow or fast tempo with the arms swaying and the hands folding in and out, gesturing with stylized movements. In the old days, *mak inang* or wet nurses were employed by the Malay aristocrats to look after the needs of their infants and young children. It is believed that they may have created the languid and easy-going dance with simple walking steps while they sang lullabies and folk songs. Although the term *mak inang* is a special reference to the lady-like qualities of the wet nurse-dancers, the dance itself eventually took the form of a folk dance when men and women danced to the tunes of *mak inang* songs.

Dancing the *inang* is relatively easy, its rhythm consisting of a variation of the 4/4 beat pattern, which is accented at the end by the gong, and it could be performed at a relatively slow or fast tempo. A fast-paced *inang* is also known as *masri*, 'with a rhythmic pattern similar to the Middle Eastern *beledi* dancing' (Nor, 2007, p. 293). *Beledi* is often spelled as *beladi, beladi*, or *balladi*, which is the most common rhythm among music used for oriental dancing that may be a generic dance of the bellydance or *raqs sharqi*. *Masri*, as a variant of *inang*, manifests the eclectic nature of the *inang* genre, embracing a rhythmic style common in Middle Eastern music. Dancers face one another while making turns and dancing in a circular path around each other. The *inang* was commonly used by dancers in the *bangsawan* Malay opera as a base for the creation of art dances that used various common objects as dance accessories, often named after dance properties. *Tarian piring* (saucer dance), *tarian lilin* (candle dance), *tarian selendang* (shawl dance), *tarian saputangan* (handkerchief dance), and *tarian kipas* (fan dance), among others, are examples of *inang* art dances that have evolved through time. Sometimes these art dances would incorporate dance movements and rhythms from *joget* and *asli* tunes.

The third repertoire of *ronggeng* is the *joget* dance, which is also referred to as *tandak*, believed to have been developed through syncretic processes that involved Portuguese and Malay influences. *Joget* or *tandak* came about with the introduction of eclectic collections of musical instruments (violin, accordion, drums) and tunes borrowed from the Portuguese in the 16th and 17th centuries. Although the performative past of *joget/tandak* is rooted in *branyo* and Kristang (old Portuguese dialect) lyrics, *branyos* has changed considerably today, 'sandwiched between latest hits and played by pop bands using electric guitar, keyboards, and trap set (replacing the old-fashion hybrid ensemble of violin and Malay *rebana*)' (Sarkissian, 2000, p. 108). However, the scarcity of information on the *branyo*, other than the studies done by Margaret Sarkissian (2000), from Malaysian and Portuguese sources suggests that it could have been a variant of an older dance and music tradition from Portugal, the *brundo* or *branle*. There are no known records of *branyo* in the Portuguese repertoire in Portugal, which may suggest that it could have been a corrupted form or mispronunciation of *brundo* or *branle*, which was a 15th-century court dance popular in the courts and countryside of the Iberian Peninsula and Italy in the 16th century. The *brundo* or *branle* was typically danced in the round, which might change into a file advancing sideways. The *branle* double and *branle* simple were in duple metre, whereas the *branle gai* was in triple metre (Nor, 1993, pp. 39–40, 49). It is possible that the early Portuguese settlers in Malacca could have performed a variant of *brundo* or *branle*, which was closely related to 6/8 dance forms such as the *tarantella* or *fandango*. Archaic *joget/tandak* shared similar performative styles of singing and exchanging of quatrains (*pantun*) between performers, which could be sung to the accompaniment of drums and gong, which provide rhythm and agogic accents. The singers-dancers dance simple steps of walking and skipping between singing the quatrains. The Malay *pantuns* are improvised stanzas similar to the *mata kantiga*, a duelling genre of exchanged quatrains sung to the accompaniment of *branyo* music. Sarkissian clarifies, 'with the exception of language, *branyo* and *mata kantiga* are – today, at least – virtually interchangeable with the Malay genres *joget* and *dondang sayang*, respectively' (2000, p. 181). The evolution of *joget* or *tandak* infused by the eclectic music of *branyo* and the stylistic singing of *mata kantiga* has made *joget* a fast-paced dance with duple and triple beat divisions enabling dancers to dance and sing in duets or in the free folk style of dancing in rows or in groups. The *joget* is known as *lagu dua* in the Sumatran *ronggeng* and was one of the main repertoires of the Malay *ronggeng* (*ronggeng melayu*) during the heyday of taxi dancing in the 1930s to 1960s. To a certain extent, *joget* is synonymous with *ronggeng*. Tan Sooi Beng clearly states:

> *Ronggeng* is a type of social dance in which mixed-sex couples dance and exchange verses to the accompaniment of a violin, accordion, one or two frame drums, and suspended gong. The tambourine and maracas are often added. It is believed that the ronggeng developed during the post-Portuguese period in Melaka, and the *ronggeng* troupes comprising performers of different ethnic origins still provide music and entertainment during Malay and Baba[4] weddings and other festivities in Malaysia . . . *ronggeng* appeals to different ethnic groups, as the repertoire is eclectic.
>
> *(Tan, 2007, p. 288)*

Pentas ronggeng (*ronggeng* dance stage) in amusement parks, and *ronggeng*-derived art dances in *bangsawan* (Malay opera) popular in the 1930s and 1940s, played important roles in popularizing *ronggeng* to urban Malayans in Malaya and Singapore. Permanent *ronggeng* dance stages staffed with taxi dancers or dance hostesses became permanent fixtures in amusement parks in Penang, Kuala Lumpur, and Singapore, whilst mobile *ronggeng* ensembles plied between villages and towns to provide social dancing at weddings and other social functions. *Bangsawan* theatre utilized *ronggeng*-derived repertoires within *bangsawan* plays or in vaudeville-like intermissions between plays called *extra-turns*. Through the growing popularity of the *bangsawan* theatre in the 1930s–1940s, dance choreographers in *bangsawan* theatres re-invented choreographies from the *ronggeng* dance cycle of *asli*, *inang*, and *joget* tunes for *bangsawan* plays or *extra-turns* (Nor, 2019, p. 54).

In the 1950s and 1960s, Malay movies and cinema networks in Malaya and Singapore presented new dance repertoires from re-invented *ronggeng*-derived art dances to serve new musical impressions, sung and danced by famous movie stars and chorus lines on silver screens. Indian movie directors who introduced Indian stories and plots into Malay movies combined newly created *ronggeng*-derived art dance with new musical arrangements and songs, thus changing the course of *ronggeng* from a social dance into derivative arts dances in the 1970s well into the 1980s and 1990s (Nor, 2019, p. 54).

Today, pockets of *ronggeng* groups performing the *ronggeng* dance cycle are still found in Kedah, Penang, and Malacca, entertaining visitors at weddings and other feasts.

> at these gatherings the *ronggeng* female entertainers dance with the male guests, and the music is provided by the *ronggeng* ensemble. Singing is done by special singers and not by the dancers themselves. Today, the *ronggeng* tunes such as the *asli*, *inang* and

joget are also performed by pop bands, the orchestra of Radio and Television Malaysia (RTM) and by symphony orchestra.

(Matusky & Tan, 2017, p. 300)

Conclusion

The 20th-century Malay social dances were conceived through the mixing of eclectic and syncretic forms and styles of dance, diachronically and synchronically, within a multicultural crucible of exchanges, influences, and dominances that were autochthonous (derived from the place it was found) and allochthonous (originating somewhere other than where it is found) in origin. Adapting to the ever-changing environment of cross-cultural interactions marked by demographic and socio-cultural shifts creating syncretic traditions through an eclectic fusion of performative traditions in a multi-cultural nation, Malay social dances continue to evolve syncretically as hybrids, autochthonously and/or as allochthonous derivatives.

Although *zapin* may be regarded as an autochthonous invention by virtue of its unique mélange of Malay music and dancing styles, its origin remains allochthonous. The symbiotic relationship of *zapin* and *zaffin* as derivatives of Islamic performing arts, connecting Hadhramaut and the Malay world or *alam melayu*, has continued to play important roles in the evolution of the *zapin* today. These two traditions were conceived through exclusively different cultural experiences with a shared history of performative aesthetics and functionalities.

Similarly, with *ronggeng*, the hybridization of autochthonously and/or allochthonous synergies is best seen in the cyclic repertoires of *asli, inang*, and *joget*, musically and in its structured movement systems. The call-and-response song quatrains, pedestrian dance movements, languid arm swaying, and flexed fingers are autochthonous in form and styles, but the eclectic mix of music repertoires, relatively faster dance movements, duple and triple rhythms, and fiesta-like performances are remnants of allochthonous vestiges.

While the *zapin* remains exclusively a social dance with Arabic-Islamic affinities, performed mostly in villages and/or in smaller towns, *ronggeng* found its niche in urban amusement parks in the first half of the 20th century, morphing into art dance from the 1980s to the 1990s. Diachronically and synchronically, the eclectic and syncretic processes continue to nurture Malay social dance to the present time.

As described in this chapter, Malay social dances have been conceived to closely intertwine with stylistic affinity with the world or *alam* of the Malays and borrowed traditions from the Middle East, Europe, and Asia. The making of Malay social dances and the consequential heritagization of constructed corporeality from multiple sources and from within their own familiar world that surrounds them transpose their corporeal traditions into a larger cultural polyglot signifying multicultural ownership. Today, these dances could be claimed by all Malaysians as their own, in spite of the peculiarities of their origins. The previously mentioned social dances epitomize the ever-changing discourses in cultural possession of intangible cultural heritage in Malaysia. Much like other art forms, Malay social dance such as the *zapin* and *ronggeng (asli, inang*, and *joget*) represent the interconnectivity of dance and its performative meanings and the socio-cultural perspective of Malay aesthetics and Malaysian idealism. In spite of the infusion of eclecticism and syncreticity, all these dances are legitimized through the stylistic codification of movements that resemble and symbolize its cultural meanings comprehended by the beholders, legitimizing eclectic and syncretic processes to uniquely create Malaysian social music and dancescapes.

Note: This is a revised and updated version of the original chapter in Mohd Anis Md Nor and Burridge, Stephanie (ed.) (2011) *Sharing Identities: Celebrating Dance in Malaysia*, Routledge, India.

Notes

1 'Performative' is both a noun and an adjective (Schechner, 2006, p. 123). The concept of the performative, which was explored by J.L. Austin in *How to Do Things with Words* (Blackburn, 2005, p. 272), delivered at Harvard University in 1955, described the state of utterance, which leads people to perform acts. In performance studies, performativity is understood to imply 'performance principle' in all aspects of social and artistic life, which is closely related to postmodernism.

2 Hadhramaut, which is today located in the People's Democratic Republic of Yemen, is made up of a valley complex in the middle of southern Arabia and is separated from the southern coast of the Arabian Peninsula by a mountain range. The location of Hadhramaut on the great trade route from South-East Asia to the Mediterranean coast has, since Roman times, also contributed to the out-migration of the Hadhramis to South-East Asia (Nor, 1993, p. 4).

3 *Taksim* is derived from the Turco-Arabic word '*taqsim*,' which means 'division' or 'sectionalization' of an improvisational musical form for a solo instrument, which depends largely on the musician's individual ability to perform to a given central tone of a *maqam*.

4 The terms *Baba* and *Nyonya* are used in reference to the Chinese (or their ancestors) who were born in Penang, Malacca, and Singapore from the Malacca period to the British rule. They are also known as Straits-born Chinese, obviously referring to the British Straits Settlement. They speak a mix of Malay and Chinese as a home language and have incorporated Chinese and Malay cultural traditions into their lifestyle (see Nor, 1993).

References

Blackburn, Simon (2005) *Oxford Dictionary of Philosophy* (second edition). Oxford: Oxford University Press.

Ho, Engseng (2006) *The Graves of Tarim: Genealogy and Mobility across the Indian Ocean*. Berkeley and Los Angeles: University of California Press.

Matusky, Patricia, & Tan, Sooi Beng (2017) *The Music of Malaysia: The Classical, Folk, and Syncretic Traditions*. London, UK: SOAS Musicology Series, Routledge.

Nor, Mohd Anis Md (2019) "Ronggeng Re-Invented: Emergence of New Repertoires from Singapore to Peninsular Malaysia," *Proceedings 5th Symposium of the ICTM Study Group on Performing Arts of Southeast Asia*. Edited by Patricia Matusky, Wayland Quintero et al. Kota Kinabalu, Sabah: Department of Sabah Museum, 54–59.

———. (2009) "Playing Is Dancing: Permissibility and Legitimacy of Malay-Islamic Structured Movement System in the Malay World of Southeast Asia," *Proceedings 25th Symposium of the ICTM Study Group on Ethnochoreology*. Edited by Mohd Anis Md Nor, Elsie Ivancich Dunin, Anne von Bibra Wharton et al. Kuala Lumpur: Cultural Centre University of Malaya and Ministry of Information, Communication and Culture of Malaysia, 185–168.

———. (2007) "Structural Constructs in Indigenous Dances in Malaysia," in *Dance Structures: Perspectives on the Analysis of Human Movement*. Edited by Adrienne L. Kaeppler and Elsie Ivancich Dunin. Budapest: Akadémiai Kiadó, 357–362.

———. (2004) "Zapin," in *The Music of Malaysia: The Classical, Folk and Syncretic Traditions*. Edited by Patricia Matusky and Tan Sooi Beng. SOAS Musicology Series. Hampshire and Burlington: Ashgate, 127–136.

———. Editor (2000) *Zapin Melayu di Nusantara* (translation *Malay Zapin of the Malay Archipelago*). Johor Bahru: Yayasan Warisan Johor.

———. (1993) *Zapin: Folk Dance of the Malay World*. Singapore: Oxford University Press.

Nor, Mohd Anis Md, & Burridge, Stephanie (2011) *Sharing Identities: Celebrating Dance in Malaysia*. New Delhi: Routledge.

Sarkissian, Margaret (2000) *D'Albuquerque's Children: Performing Tradition in Malaysia's Portuguese Settlement*. Chicago and London: The University of Chicago Press.

Schechner, Richard (2006) *Performance Studies: An Introduction*. New York and London: Routledge.

Tan, Sooi Beng (2007) "From Folk to National Popular Music: Recreating Ronggeng in Malaysia," *Journal of Musicological Research*, 24:3, 287–307.

11

TRANSNATIONALISM AND THE WANING OF THE NATION-STATE FOR MALAYSIAN CONTEMPORARY DANCE CHOREOGRAPHERS

Bilqis Hijjas

When I first wrote this chapter in 2010, Low Shee Hoe, a choreographer, dancer, and lighting and graphic designer from Malaysia, was working as a lighting technician for a mainland Chinese dance company on tour in Europe. At the same time, his dancers were performing a work that he choreographed in a theatre in Kuala Lumpur. Meanwhile, dancer-choreographer Suhaili Micheline was engaged in a short residency in Kobe, Japan, shivering in the cold and struggling to learn enough Japanese to find her way around. During her absence, the classes she usually taught at the National Arts Academy in Kuala Lumpur and at her suburban ballet studio were being taken over by her colleagues. Far from home, both Shee Hoe and Suhaili were both rigorously engaged in furthering their dance careers to enable them to realize their goals here in Malaysia. Their strategic negotiation of the various social, economic, and political pressures that come to bear on their paths towards creativity and performance made Suhaili and Shee Hoe useful illustrations of the transnationalist trend which increasingly marks the work of contemporary choreographers from Malaysia.

The bulk of significant Malaysian contemporary choreographers in earlier generations were nurtured overseas. Some of them travelled to the global capitals of contemporary dance; others searched for their cultural roots in the homelands of their diasporas. A partial list of the generation of established artists who received their most impactful dance education or first significant experience of dance employment overseas includes Ramli Ibrahim, Marion D'Cruz, Joseph Gonzales, Lee Lee Lan, Mew Chang Tsing, Choo Tee Kuang, Anthony Meh, Aman Yap, Suhaimi Magi, and Vincent Tan, with Umesh Shetty, Aida Redza, and Michael Voon within the subsequent generation. A number of dancer-choreographers self-identifying as Malaysians continue to do the greater part of their dance work overseas, including Mavin Khoo, Mei-Yin Ng, and Lena Ng, returning home occasionally or maintaining partial residence in Malaysia.

For those who do return to continue their dance career in Malaysia permanently, no small part of their subsequent local success can be attributed to their overseas experience and especially to the social fetishization of their return. The cultural cringe that persists after decolonization makes international credentials a requirement for local success. Artists

DOI: 10.4324/9781003160007-19

who are able to establish themselves overseas, surviving in what are seen as more cutthroat and admirable cultural environments, are guaranteed a warm reception on their return, and the event of the return itself acquires mythic status. It is often couched in self-sacrificial and nationalistic terms, loaded with the desire to share the benefits of an overseas education with a local population that is, by implication, backward and uncultivated. In reality, returning from overseas is often a choice undertaken under compounding pressures, such as economic need, familial ties, the inability to secure a visa or permanent residence overseas, the demands of raising a family, and the awareness of greater scope or opportunity in the home country. Nevertheless, the rhetoric of the personal contribution to the Malaysian nationalist project has been an enduring feature, especially among the early generations of contemporary dancers.

Within contemporary dance, and the arts as a whole in Malaysia, the post-independence challenges of being a united nation-state and of creating an arts vocabulary suitable for all Malaysians have been stimulating concerns for many artists. Ramli Ibrahim claims,

> There is a palpable awareness among Malaysian contemporary choreographers that modern ideas . . . can and should take into account the rich and varied indigenous traditions and contemporary culture [and that work within the last 20 years has] given birth to a distinctive Malaysian identity in contemporary dance.[1]

While I agree that awareness of the nationalist project has indeed been palpable, I hesitate to claim that this project has arrived at a successful resolution. Mew Chang Tsing's *Lady White Snake* project grappled repeatedly with the difficulty of

> marrying and fusing our different dancing backgrounds . . . not traditional poses stolen from a variety of styles and pieced together in a forced and confused fashion, but instead . . . the emergence of a completely new cultural identity, one distinctly Malaysian.[2]

The difficulty of avoiding simplistic pastiche collided with the imperative to resist cultural imperialism. Ramli Ibrahim maintains that the works of modern Malaysian choreographers 'bear little resemblance to, and are unaffected by, the Euro-American idiom',[3] but Marion D'Cruz accedes that, within international collaborations at least, the risk continually exists that these attempts 'allow some collaborators to colonize others . . . a most disempowering experience'.[4] The turn of the millennium, however, has marked a period of distinct disruption worldwide from which Malaysia has not been exempt. The irresistible spread of transnational capitalism and global culture, experienced in Malaysia as the currency crisis of 1997, and increasing access to information through the Internet, which led not indirectly to the political upset of the 2008 elections, as well as the availability of cheap international flights through private budget airlines while national airlines have declined, have brought Malaysia into different relations with its former colonialist powers, diasporic homelands, and regional neighbours.

In its present incarnation, Malaysia's nationalist discourse has often resulted in the outflow of artistic talent rather than its return. As the current administration tries to cling to power, it has reasserted the importance of affirmative action for the ethnic Malay community instituted under the New Economic Plan (2010), which many urban Malaysians now find outmoded. A recent resurgence in Malay nationalism has led to the concept of *Ketuanan Melayu*, or Malay supremacy, being declared sacrosanct and above discussion. The importance of defining who deserves the title of *bumiputra* (son of the soil) in order to enjoy political and economic advantages has fostered a system of racial categorization in which 'the successful continuation of this nation-building project depends . . . on the maintenance of fixed categories of race,' as Susan Philip explains in her article on Malaysian playwright K.S. Maniam's work. Philip goes on to illustrate how the reductivist effect of such categorization does violence not only to racial hybrids, such as those who have a parent from two different 'races', but also to cultural hybrids, in which she places the entire diasporic non-*bumiputra* population. 'Each race is yoked, willy-nilly, to a particular culture that is based on the culture of the original homeland . . . thus narrowing and essentializing the individual's cultural space'.[5] The lived reality of this restrictive construct is somewhat different. There is a certain amount of wiggle room, in which individuals assert their agency within the interstices of a nationalist concept imperfectly implemented. Philip recognizes this flexibility, drawing upon the concepts of Bakhtin and Gerald Baumann to observe that 'The lived reality at the demotic [grassroots] level is of organic hybridity',[6] which has led to the emergence of what K.S. Maniam himself dubs the 'new diasporic man', one who is aware not only of his own culture but also of the cultures around him and who occupies multiple cultural as well as imaginative spaces.

While such creative negotiation of Malaysia's cultural policy is possible, it is an exhaustive option, which requires a continual acknowledgement of, if not a surrender to, the essentializing discourse which undermines the claim to citizenship of much of the population. May Joseph[7] stresses the performative nature of citizenship rather than its definition as mere legal entitlement, that citizenship is continually enacted and expressed. 'Citizenship is not organic,' Joseph writes, 'but must be acquired through public and psychic participation. Citizenship is an ambiguous process vulnerable to changes in government and policy. The citizen and its vehicle, citizenship, are unstable sites.' In countries like Malaysia, characterized by the presence of large diasporic populations, there exists a 'nomadic, conditional citizenship related to histories of migrancy and the tenuous status of immigrants' in which members of diasporic communities are marked out as 'inauthentic citizens'.

If citizenship is indeed performative and constituted by a host of 'anxious enactments',[8] then how much more anxiety must be withstood by those whose career and identity revolve around the conscious production of performance? The effect of recent political changes on Malaysia's national cultural policy has been one of immobilization. The government arts ministry continues to foreground Malay cultural events at the expense of the rest of the community. The only government scheme of ad hoc grant applications to non-government artistic organizations and individuals has been frozen. In this atmosphere, it is unsurprising that many young Malaysian artists quit the country, either in an attempt to escape the psychological battery of finding themselves second-class citizens or in pursuit of greener economic and artistic pastures.

Taking Suhaili and Shee Hoe as examples, I suggest that for the younger generation of Malaysian contemporary dancers and choreographers, while the prevalence of overseas training continues, the hegemonic dominance of the nationalist project has waned. While these two dancer-choreographers may not be representative of their generation in Malaysian contemporary dance, they usefully illustrate the diversity of responses to recent globalizing trends. Through their personal transnationalist experiences, these younger artists multiply the trajectories of their identities both outwardly and inwardly, upsetting the binaries of 'us and them', and in the end disrupting the very concept of 'Malaysian contemporary dance'.

A sixth-generation Malaysian-Chinese, Low Shee Hoe grew up in a Chinese-speaking working-class family in Kajang, a satellite town of Kuala Lumpur which is predominantly Chinese. He attended Chinese primary and secondary schools, and until he was 18, his only knowledge of the other races in Malaysia had come through his school textbooks. When he went to university, he was startled and intrigued by 'how many Malays there were in the world'. He cultivated Malay friends and learned about their culture, even accompanying them in their Ramadan fast. Although his Chinese family and friends were deeply disapproving of this newfound interest, Shee Hoe was entranced by the cultural differences he was discovering.

However, it was not until he left on a scholarship to study at the Hong Kong Academy for Performing Arts (HKAPA) that he first consciously identified himself as Malaysian; up until then, he had thought of himself as Chinese. His experience of living abroad induced a period of questioning and doubt about his own identity that he has been unable to conquer. Surrounded by other Chinese from Hong Kong, Macau, Taiwan, and mainland China, what was there, he felt, to indicate that he was Malaysian? And yet he knew himself to be culturally distinct.

These thoughts soon crystallized into an experimental theatre work he created for an assignment at HKAPA. Entitled *Death*, following a poem by a nationalist Chinese poet, the short work combined lighting design, movement, sound design, and multimedia to question the basis of identity. Upon entry, the audience was requested to relinquish their identity cards – those who refused were refused entry. The identity cards were quickly scanned, and later during the work, images of them were projected onto the back wall of the venue. At the end of the work, the identity cards fell in a cascade from the ceiling, each wrapped in a medical bag like a piece of evidence. The audience was forced to cross a mess of fake blood and sort through the pile of cards to recover their own. With this work, Shee Hoe wanted to create an atmosphere of surveillance and authoritarianism and also to goad the audience into wondering at what cheap price they were willing to hand over their 'identities'. If they had lost their cards, who were they and how would they prove it?

In retrospect, Shee Hoe feels that, of all his work, this one had the most potential. But he would be uncertain about restaging it in Malaysia because he feels that to non-Chinese Malaysian audiences, and especially Malay audiences, it would come across as a typically Malaysian-Chinese demand for equal rights and equal citizenship and that this antagonistic approach would be unnecessarily alienating. He feels that he would have to couch it clearly in terms of criticism of government policy rather than Malaysian society as a whole but that this change, in addition to potentially landing him in detention for questioning Malay supremacy, would render the entire work too didactic. So Shee Hoe finds himself in the ironic position of having made a work about Malaysian-Chinese identity which plays rather better to an international Chinese audience than to a specifically Malaysian one.

And yet Shee Hoe's audience of choice is Malaysian. While he recognizes that much of its audiences are less sophisticated than those he might find in Hong Kong, and that he must make his work more accessible and legible as a result, he feels that Malaysia needs to see his work. He acknowledges that performing to Hong Kong audiences was more nerve wracking, but it is not fear of failure that motivates him to return; he could find constant employment in Hong Kong, Taiwan, or mainland China. But he feels that he has definite contributions that he can make to the Malaysian arts scene, including his proficiency with contact improvisation technique, multimedia work, and concept development. He also feels that he has a responsibility to share the vision of the dance industry that he witnessed overseas, in which people can be employed within the dance industry not just as dancers and choreographers and in which full-time dedication towards the creation of art is possible.

In addition to being able to make contributions to Malaysia, he feels that there is more that can be wrested from Malaysia. Malaysian artists, in his opinion, are more creative, because they are hungrier for art. Echoing the nationalist preoccupations of preceding generations, Shee Hoe wants to be a part of developing a truly Malaysian art form which is not merely a pastiche of traditional forms. But he realizes that this vision may be slow in coming – that Malaysia cannot rush into creating the 'cultural memory' which it has taken China thousands of years to accrete.

Suhaili Micheline's experience of citizenship and nomadism is somewhat less fraught but also more rootless. Suhaili grew up in her mother's suburban dance studio, learning ballet and bringing home a host of local ballet awards. After spending four years in Melbourne, Australia, earning her bachelor's degree in contemporary dance largely founded upon release technique, Suhaili returned to Malaysia to take up the reins at the family dance studio and also to perform and choreograph for a wider audience. During the last few years, she has picked up a slew of awards, including Most Promising Artist of the BOH Cameronian Arts Awards in 2008, Best Choreographer and Female Dancer at Short+Sweet Dance 2009, and a finalist position in the first season of Malaysia's version of the television competition series *So You Think You Can Dance?* With hard work, a technically disciplined body and choreography distinguished by a quirky sense of humour, Suhaili has quickly found local success, but it has left her largely disappointed and demoralized. Judging it against her experience of performing in Melbourne, she feels that her talent and ability remain largely untapped and that the Malaysian dance community encourages laziness and a false sense of importance. She yearns to continue dancing overseas and to finally secure a full-time contract with a reputable international company, both to challenge herself and to ensure that her local reputation is based upon standards of international quality.

With a Chinese mother and a Malay father, Suhaili epitomizes Philip's 'racial hybrid' in contrast to Shee Hoe's 'cultural hybrid'. Rather than feeling, as Shee Hoe does, that her claim to citizenship is in doubt, Suhaili is aggravated by the constant association drawn between her racial hybridity with her personal ability. At first glance, she is often taken as Chinese, until people hear her name, and this misunderstanding has been the site of much culturally loaded discourse upon the origins of her talent and success. She recalls an encounter with a government official while she was organizing a workshop between Australian and Malaysian dance students. 'You're Malay, are you?' the civil servant asked. 'Eh, you're very good, aren't you?' The nudge-nudge wink-wink quality in the exchange left her feeling demeaned, as if the civil servant were making her complicit in a supremacist agenda to demonstrate Malay supremacy. In another case, a journalist covering the *So You Think You Can Dance?* series asked Suhaili why she did not make her Malay identity more explicit during the show, which would have made her more attractive to the Malay voting audience. While Suhaili is disgusted and disappointed by the suggestions that she should exploit her Malay identity in exchange for favours, she suffers under the fear that others might believe her success to be attributable merely to her politically favourable status rather than reflecting her personal talent. She is also aware that many consider her physical talent and work ethic to be directly inherited from her Chinese mother rather than her Malay father, in reference to the popular racial stereotypes of industriousness versus laziness.

Despite the fact that Suhaili has never felt ambiguity about her Malaysian identity, or perhaps because of her confidence in it, she is unwilling to foreground her citizenship in an international context, fearing in some cases that her audience will underestimate her, as once in Australia, her fellow students assumed she could not speak English. During her overseas engagements, she claims she does not feel any pressure to represent Malaysia (which is the experience of many who live and work in exile, feeling themselves more representative and more nationalist abroad than at home). Suhaili thinks rather that she must represent herself first and second represent a broadly defined non-Caucasian minority culture. Neither does Suhaili feel the need to directly engage with the fraught struggles surrounding the nationalist/identity rhetoric in Malaysia, preferring to think of herself as an entertainer during a time of social trial, in the tradition of the Hollywood musicals that leavened the awfulness of daily existence during World War II.

Interestingly, it is this unwillingness to play to established stereotypes, both nationally and internationally, which has in some cases (including perhaps *So You Think You Can Dance*) cost Suhaili the crown. In 2009, she created the short work *Nerds Gone Nuts* for the Short+Sweet Dance Festival at Kuala Lumpur Performing Arts Centre. Based on the Short+Sweet Theatre Festival, which originally premiered in Australia, the winner of the Malaysian version was promised an appearance at a subsequent Short+Sweet dance festival in Australia. From opening night, *Nerds Gone Nuts* was a crowd favourite, with its studiously deadpan humour performed by dancers surrounded by an exuberant collection of globally recognizable nerdy paraphernalia: thick-rimmed glasses, Rubik's cubes, Nintendo PS computer games, and Transformers masks. The work was favourably received by the local judges, who eventually awarded Suhaili the titles of Best Choreographer and Best Overall Production. Despite the accolades, the Australian Short+Sweet representatives eventually selected a competing piece for performance in Australia, much to the outrage of the local judges. *Nerds Gone Nuts*, it vaguely emerged, was deemed too globalist, or perhaps even too Australian (marked by Suhaili's Melbourne-cultivated sense of humour and style), to represent Malaysia to an Australian audience. While this decision came as a deep disappointment to Suhaili, who had deliberately constructed the work with an Australian, and even a Singaporean or Thai, audience in mind, she refused to accept that in all likelihood her work had been rejected because it refused to pander to an essentialist view of Malaysian art as either highly politicized or exotic.

Suhaili's desire to perform on the main stage rather than one demarcated for minorities, and her unwillingness to capitalize upon the performance of difference, have characterized her efforts to assert her identity in both a national and international context. Her occasional trips overseas to participate in small international events for dancer-choreographers may well precede a more extended period of leave.

Between these two young Malaysian dancer-choreographers, there is a common positive approach towards the adoption of international techniques and methods. Shee Hoe, for example, feels that contact improvisation is not an American technique loaded with American preconceptions about the contraction of space between bodies and the mutual responsibilities of weight sharing but a neutral technique which has found global acceptance and which can be usefully employed in Malaysian work if it serves to elucidate a necessary concept. 'This is no conflict,' he said, 'The concept is the most important thing.' Suhaili, meanwhile, does not necessarily consider her audience Malaysian during the creation of her work. Rather she imagines how her mentors or idols in Australia might approach certain tasks and challenges – 'What would Sue Healey do? Or Gary Stewart?' – in order to exert a type of international quality control upon her own work. Her deepest concerns about being a 'Malay' woman performing ballet revolve around disapproval of her flesh-baring tutu in a recent production at the government-run National Theatre. On hearing Ananya Chatterjea, on a recent visit to Malaysia, passionately assert that non-White dancers should not perform in cultural imperialistic White dance forms, Suhaili's response was 'That's ridiculous!' In her opinion, not only is a dancer not responsible for the particular dance forms to which he or she is first exposed and which remain deeply imprinted on the body, but a choreographer should also not censor what kinds of movement he or she finds inspiring.

At first glance, these approaches might suggest a lack of political awareness among the younger generation of Malaysian choreographers. Marion D'Cruz observes a similar occurrence during a collaborative project between Malaysian and Japanese dancers:

> The historical relationship between Japan and Malaysia loomed in the consciousness of both the older members and the highly politicized members of the group [but] the younger members had no knowledge, nor serious concerns for this historical past.[9]

Being aware of something, however, is distinct from thinking that it is applicable. Both Suhaili and Shee Hoe demonstrate a familiarity with the concept that non-Malaysian movement is inappropriate for Malaysians, but they reject it, and not, I would argue, through mere comfort and laziness but with a pragmatic understanding that current trends of globalization in Malaysia prevent an easy discrimination between the previously hegemonic binaries of East and West, Us and Them, colonizer and colonized, and also that the movement vocabularies in which their bodies 'naturally' speak make them living, breathing, dancing examples of inescapable hybridity.

Together with this easy cultural eclecticism, both Suhaili and Shee Hoe demonstrate an ability to transcend national boundaries in the performance of their work. They naturally accept the challenges of performing to overseas audiences, which brings with it the need to address and engage with entirely different world views with different interests and preoccupations. Their ability to participate in what is rapidly becoming a global art market, simultaneously present at home and abroad, is what marks them as emerging transnational agents. Arjun Appadurai, an early scholar of transnational cultural studies, emphasizes the ways in which global cultural flows occur along multiple dimensions, including through media, technology, finance, ideologies, and, most importantly, through the movement of human bodies themselves. While celebrating the possibilities that transnationalism offers for the construction of new identities that transcend borders, Appadurai also points out that these developments are inversely related to the integrity of the nation-state whose very existence is derived from the maintenance and defence of national borders.[10]

Referencing his examination of intellectuals and artists in China, Sheldon Hsiao-Peng Lu explains the mechanism by which the decline of the nation-state, globalization, transnationalism, and postmodernity are linked within the body of the Third World artist:

> Throughout the twentieth century, and up until the last decade, Third World intellectuals and artists were very much national intellectuals and artists, for they were more or less anchored in an organic relation with the nation-state. Although they may have been dissatisfied with, or even persecuted by, the nation-state, it was still

> the nation-state that provided them with a sense of identity and a platform for expression. With the advent of globalization and commercialization at the fin de siècle, the nation-state is no longer the principal site of activity for Third World intellectuals and artists . . . they participate in critical activities and academic politics in transnational settings, thus skipping the link of the nation. . . . Third World artists step directly onto the supranational scene and operate in the global art market. . . . Yet paradoxically, their international credibility rests on an indigenous appeal, a cultural localism, a certain Chineseness. This double orientation of the local and the global that besets intellectuals and artists, with all its contradictions, points to a condition of postmodern transnationalism.[11]

It is worthwhile extricating several strands of Lu's argument in relation to our Malaysian examples. First, many young Malaysian contemporary dance artists, such as Shee Hoe, continue to be motivated by their relationship with the Malaysian nation-state, even while operating on the 'supranational' level. Suhaili continues to struggle with the ramifications of comments made about her success and ability, comments which derive their stinging power from the system of racial categorization that has been a cornerstone of the Malaysian nationalist project. The nation-state may no longer be the only unit of analysis, but in Malaysia, it is far from dead; while its effect on lived reality may be complicated by globalization, it is certainly not negated.

Second, acceptance of the opportunities of transnationalism, the ability to 'step directly onto the supranational scene', must not be accompanied by lack of awareness of the continuing asymmetrical relations of power which exist in the transnational postmodern condition. In the words of radical Chicano performance artist and writer Guillermo Gómez-Peña, for those on the benign side of globalization, 'We are now allegedly installed in a fully globalized, post-racial, post-racist, post-sexist, post-ideological, post-civil rights era, and anyone who thinks otherwise is clearly out of touch with the times',[12] but for those on the 'dark side' of globalization, the reality is quite the opposite. While the global art market provides new opportunities, these opportunities are not always empowering ones. For Shee Hoe, for example, the chance to work full-time with Guangzhou Modern Dance Company in its European tour is only possible if he leverages his most marketable ability – lighting design – while temporarily subsuming his desire to dance and choreograph. For Suhaili, achieving the contract of her dreams in an internationally respected dance company would also mean renouncing her developing role as choreographer. She would risk becoming a 'hired dancer', adaptable to whatever style or genre is required, as a 'predominantly commodified identity',[13] or a 'mute colonized female body'.[14] Granted, while trends dictate that contemporary dance choreographers demand more and more input from dancers, this risk of being muted is minimized. On the other hand, Shee Hoe may find, as Suhaili already has, that as long as 'their international credibility rests on an indigenous appeal', Orientalist and exploitative positions familiar from the past will continue to shape the international opportunities that they encounter.

Finally, as Lu points out elsewhere in his article, postmodernism as experienced in the Third World nation is a messy and unevenly distributed process, 'a palimpsest of nonsynchronous, emergent and residual formations'.[15] While wielding a postmodern scepticism towards metanarratives, it would be erroneous to predict that the practice of contemporary dance in Malaysia will increasingly move towards a post-national transnationalism. In addition to increasing influence from dancer-choreographers like Suhaili and Shee Hoe, the next decade in Malaysia will also probably see the rise of dance graduates from ASWARA, the National Arts Academy. Trained entirely in Malaysia in a combination of traditional and contemporary styles, many of them lacking any previous dance training, ASWARA graduates may finally embody the nationalists' hopes for a truly Malaysian form of identity in dance. A number of recent ASWARA graduates have, however, already demonstrated a tendency to go overseas in search of a deeper dance education experience, their ASWARA credentials providing them with a springboard to access international academic structures. The continuing paucity of government support in Malaysia will probably continue to make international travel an appealing option for those with the talent and initiative.

What does this mean for Malaysian dance as a whole? If success within the Malaysian contemporary dance movement thus far has depended upon engendering a seamless and organic fusion of the diversity of Malaysia's cultural traditions, and if Malaysia's uniqueness in the world, therefore, is situated in our remarkable cross-cultural syncretic facility, then it would seem that we are ideally placed to take advantage of the opportunities offered by postmodernity and transnationalism to produce dance characterized by borderless eclecticism. And yet, at the exact moment of nationalist pride at discovering ourselves at the forefront of global movements, do we not also suddenly find the ground giving way beneath our feet, the very notion of a specifically Malaysian identity having dissolved into nothingness? Malaysian artists simultaneously marketing themselves at home and abroad will become Southeast Asian artists, subsequently Asian artists, and eventually global artists with no fixed home address – whither, then, our identities?

These are questions that as yet have no answers, as we still struggle to find our feet amidst what Gómez-Peňa calls the 'constantly shifting fault lines in an ever-fluctuating landscape'.[16] Dance scholars in other Asian nations[17] have taken comfort in interculturalism, Ortiz's theory that stresses the vitality and the 'local genius' of societies acquiring foreign cultural material, rendering globalization merely an 'exciting adventure'. More pragmatic in confronting the threats of enforced homogeneity through cocacolonization, Cheryl Stock suggests that

> Perhaps the myriad of subtle differences which mark each dancing body as unique . . . can be more appositely described as 'accented' rather than hybridised. . . . How we choose to accent our multiple identities is how we present our art and selves.[18]

Stock's admonition to pay attention to each dancing body reminds us not to beat homogenizing globalization to the finish line by eliding the specific differences that exist between individuals. As dance writers, the meanings we construct in our consumption of contemporary dance are also markers of specific locality by which, in turn, the subjects of our analysis may be known. Arjun Appadurai, the pioneering social theorist of the transnational public sphere, writes that when the nation-state faces transnational destabilization, then locality itself seems to have lost its moorings. Locality is a fragile social achievement, according to Appadurai, a relational and contextual quality constituted partly by a sense of social immediacy and requiring effort in its maintenance.[19] So if we may, to extrapolate from Appadurai's words, be each other's 'home in the world', then we must pay careful attention to the specificities of the individual artists who make up our communities as we define them.

So we return to Suhaili, who is performing a solo work in a theatre in Kobe. Having laid out a selection of her own shoes, from gold lamé sneakers to crimson stilettos, symbolizing walking in the footsteps of famous female predecessors, she is now thinking about Benazir Bhutto. Wearing a loose headscarf, she sits darning a hand-embroidered shoe. With little scooting steps, she revolves on her stool, then struggles to rise. At one point her raised arm with the needle in it resembles a power salute. At another point it suggests a Muslim call to prayer. In the original version of the work, the performer who developed the movements in this section was untrained in dance, and Suhaili has enthusiastically embraced this quality in the movement. Later, she is surprised that some Japanese members of the audience have been moved to tears.

Meanwhile, Shee Hoe's dancers in Kuala Lumpur are launching into his work, apparently unconcerned by his absence. Within a pattern of white paper suitcases meticulously arranged on the stage, groups of dancers take turns repeating a movement phrase presented at different angles to the audience. The phrase is characterized by huge swirling twists, punctuated with sudden graceful inversions and a single staccato jutting hip. With every turn, the dancers seem in danger of sending the fragile white suitcases flying, but they never do. Watching it, I am reminded of Shee Hoe's own space-eating movement quality, now multiplied onto ten other bodies, and how characteristic it is for him as a stage designer to use props or lights to create restrictive but frail boundaries within which the human body seems miraculously contained.

Reviewing this chapter ten years after it was written, much has changed for the Malaysian contemporary dance community, yet much has remained the same. A new generation of local dance artists, including Lee Ren Xin, Lau Beh Chin, Khairul Mokhtar, Ng Xinying, Faillul Adam, and Murni Omar, has returned clutching their overseas qualifications and hoping to make good at home. There has been an uptick in the visibility of Malaysian productions in arts festivals and events overseas, courtesy of works by Five Arts Centre, TerryandTheCuz, and Inner Space Dance Company. The advent of CENDANA, the Cultural Economy Development Agency, has made more federal funding for contemporary arts available and has introduced the global discourse of the creative industries into the local context.

Yet some developments that appeared substantial may turn out to be fleeting. The general election of 2018, which ushered in a new party in power for the first time in Malaysia's 60-year history, was followed by a backdoor coup in early 2020 which returned the old government to power. The rise of authoritarian populism around the world is mirrored in Malaysia by ultra-nationalist support for the disgraced former prime minister, who is implicated in the theft of billions from the national coffers. The result is increasingly polarized communal politics, distrust of immigrant communities, and a rise in extremist Islam.[20]

The evolution of the dance careers of the two individuals highlighted in this chapter illustrates these changes on a micro scale. Low Shee Hoe has remained overseas, serving as resident lighting designer of Guangdong Modern Dance Company in Guangzhou, China, since 2009. As a showpiece of China's soft diplomacy, it offers unparalleled reach; Shee Hoe recently travelled with the company on an extensive tour to Europe and North America. He retains connections with Malaysia, occasionally returning to visit family and once, in 2018, as technical director and lighting designer for the major theatrical production *Bisikan Monsoon* (directed by Chang Wei Loy, another Malaysian-born Chinese theatre maker who has found a professional niche for himself in Taiwan). While Shee Hoe enjoys a stable job and global scope for his talents as a lighting designer, his career as a choreographer and as a member of the Malaysian arts community has largely evaporated.

Despite her aspirations to seek employment abroad in 2009, Suhaili Micheline has remained in Malaysia. Now hailed as an entrepreneur[21] and a minor local celebrity, she continues to helm her family's chain of dance studios while adopting positions as brand ambassador for local and international fitness products and as a partner in a new fitness studio. Although Suhaili's choreography often takes a back seat to her other work, she continues to produce new dances. She recently presented a half-evening work, *Pendatang Pampers*, alongside a work by fellow returnee Kenny Shim, in January 2020. In contrast to Shim's delocalized cosmopolitan style, Suhaili's work exhibited a trenchant Malaysianness, using humour (even toilet humour), as well as her now-trademark quirky sensibility and cartoonish costumes, to poke fun at the dignity of pompous politicians.

The 2020 COVID-19 pandemic has brought issues of localism and transnationalism into sharp relief. Nationally coordinated responses to the pandemic have shunted nationalism and the nation state to the fore: borders have slammed shut, international travel is curtailed, and foreigners are viewed with apprehension. Yet while physically our world has been shrunk, in other ways it has been broadened. The pivot to digital prompted by widespread lockdowns means that Malaysian artists have been forced onto a global playing field, where they must compete for local eyeballs against world-class players who have better resources and more marketing savvy. Like the old model of the transnationalist art industry, which involved transporting bodies and physical art works across state lines, the new model offers its own challenges as well as opportunities. It too reproduces existing geopolitical inequalities while extending new promises of liberty and identity for art makers. How dance will survive its transmutation to digital media, how Malaysian dancers will navigate through this virtual space, and whether they remain tied to the project of the Malaysian nation state remains to be seen.

Note: This is a revised and updated version of the original chapter in Mohd Anis Md Nor and Burridge, Stephanie (ed.) (2011) *Sharing Identities: Celebrating Dance in Malaysia*, Routledge, India.

Notes

1 Ramli Ibrahim (2003: 23).
2 Murallitharan (2000: 6).
3 Ramli Ibrahim (2003: 29).
4 Marion D'Cruz (2003: 77).
5 Philip (2004: 179).
6 Philip (2004: 179).
7 May Joseph (1999: 2).
8 Ibid (1999: 2–5).
9 Marion D'Cruz (2003: 86).
10 Appadurai (1996: 33, 158).
11 Sheldon Hsiao-Peng Lu (1997: 96).
12 Guillermo Gómez-Peña (2001: 12).
13 Stock (2005: 23).
14 Desmond (2001: 41).
15 Lu (1997: 66).
16 Gómez-Peña (2001: 7).

17 Carino (2005: 101) and Chen (2000).
18 Stock (2005: 23–24).
19 Appadurai (1996: 178–179).
20 *The Economist* (2016).
21 Yim (2019).

Author's comment

Insight into the lives and methods of Low Shee Hoe and Suhaili Micheline binti Ahmad Kamil is derived from years of friendship and observation of their work and also significantly from interviews conducted on 25 November 2009 and 11 December 2009, respectively.

References

Appadurai, A. (1996). *Modernity at Large: Cultural Dimensions of Globalization*. Minneapolis: University of Minnesota Press.

Carino, C. (2005). Globalization's Exciting Adventure: Intercultural Contemporary Dance Production (Singapore). In M. Anis & R. Murugappan (Eds.), *Global and Local Dance in Performance*. Kuala Lumpur: Cultural Centre University of Malaya and Ministry of Culture, Arts and Heritage Malaysia.

Chen, Y.P. (2000). Transculturating Modern Dance: Voices of Innovation and Protest in the Post-Cloud Gate Taiwan. In M. Anis (Ed.), *Asian Dance: Voice of the Millennium*. Kuala Lumpur: Asia Pacific Dance Research Society and Cultural Centre University of Malaya.

D'Cruz, M. (2003). Collaborative Efforts, Experimentations and Resolutions: The Way to Contemporary Dance. In M. Anis (Ed.), *Diversity in Motion*. Kuala Lumpur: MyDance Alliance and Cultural Centre University of Malaya.

Desmond, J. (2001). Dancing Out the Difference: Cultural Imperialism and Ruth St. Denis's 'Radha' of 1906. In A. Dils and A. C. Albright (Eds.), *Moving History/Dancing Cultures*. Middletown: Wesleyan University Press.

Gómez-Peña, G. (2001). The New Global Culture: Somewhere between Corporate Multiculturalism and the Mainstream Bizarre (a Border Perspective). *TDR 45*(1), 7–30.

Joseph, M. (1999). *Nomadic Identities: The Performance of Citizenship*. Public Worlds, No. 5. Minneapolis: U of Minnesota Press.

Lu, S.H.P. (1997). Global Post Modernization: The Intellectual, the Artis, and China's Condition. *Boundary 2 24*(3), 65–97.

Murallitharan, M. (2000, September 8). A Different Lady White Snake. *New Straits Times*.

Philip, S. (2004). Diasporic Spaces in K. S. Maniam's "The Sandpit: Womensis". *Asian Theatre Journal 12*(2), 177–186.

Ramli Ibrahim. (2003). Indigenous Ideas and Contemporary Fusions: The Making of Malaysian Contemporary Modern Dance. In M. Anis (Ed.), *Diversity in Motion*. Kuala Lumpur: MyDance Alliance and Cultural Centre University of Malaya.

Stock, C. (2005). The Cocacolonisation of Difference: Homogenized Diversity in the 21st Century Cultural Practice. In M. Anis & R. Murugappan (Eds.), *Global and Local Dance in Performance*. Kuala Lumpur: Cultural Centre University of Malaya and Ministry of Culture, Arts and Heritage Malaysia.

The Economist. (2016, March 3). The Najib Effect.

Yim, J. (2019, December 13). Chasing Dreams with Suhaili Micheline, Marini Ramlan, and Alena Murang. *Prestige Malaysia*.

Philippines

12

THE FOUNDATION OF LANGUAGE

New Filipino dance lexicons from Eisa Jocson

Vanini Belarmino

Witnessing Philippine dance from the sidelines

My understanding and love of dance was founded on my early years as a young cultural worker at the Cultural Center of the Philippines (CCP)[1] in the 90s. Spending close to a decade of many long hours observing, listening, watching, and working with Philippine dance icons – from ballet to folk dance at rehearsal halls, studios, and the main theatre stage – offered and allowed me to have a special view on lived dance history from behind the curtains. Nestor O. Jardin, then the CCP Performing Arts Department director who later became the Centre's artistic director and president, decided to take me under his wing while I was still at university to work on various projects, with roles from production to stage management, press relations and marketing to project coordination. This meant having direct access to artistic directors and company members of the CCP Resident Dance Companies[2] like Ballet Philippines (BP),[3] Philippine Ballet Theatre (PBT), Bayanihan Philippine National Folk-Dance Company, and Ramon Obusan Folkloric Group (ROFG). The immediate environment where I operated was made up of choreographers, dancers, and artists, shaping and defining Philippine dance history. This included people integral to the ecosystem: the late Philippine National Artists for Dance,[4] Lucresia 'Mommy' Urtula and Ramon Obusan, founding directors of Bayanihan and ROFG, respectively; PBT's Gener Caringal; BP's Agnes Locsin. There were also CCP's regular collaborators, who ranged from the likes of Corazon Inigo, long-time artistic director of University of the Philippines Filipiniana Dance Group, and Basilio Esteban Villaruz, dance historian, critic, and academic, who worked on aspects from staging productions to authoring books and building the video archives.

It was there that I organized the 124 company members and staff of the Bavarian National Ballet in 1994 for the monumental production of *Romeo and Juliet* featuring Philippine pride and the German state ballet's principal ballerina, Anna Villadolid. The presentation of an international touring production of such scale was part of CCP's commitment to presenting works by Filipino artists and so-called world-class productions. Making this production happen opened my eyes to new ways of working. What was especially valuable

DOI: 10.4324/9781003160007-21

was seeing how a dance company could be in a position to set the conditions by which its work is presented.

From being in close proximity with colleagues who regularly travelled abroad with CCP resident companies on cultural diplomacy missions, I learnt that there is a need to arrange for 'host families' on some occasions to economize resources. Joining an international tour was considered a privilege and held a deep meaning of representing the country.

In 1995, I landed a job fulfilling multiple roles with BP. I was exposed to BP's suite of programming – from classical ballet to neo-Filipino productions – and also gained perspectives on how a dance company functions, tours, and develops choreographic works. Among my responsibilities was to stage-manage all the shows. This involved creating special notation for the choreography and blocking of Agnes Locsin, the company's artistic director. This process gave me an intimate understanding of how movement is planned for and implemented and how studies are made. It was then that I began memorizing choreography, studying the physicality of dancers, learning about the natural capacities and limitations of the body and the pulse of the music.

One of Locsin's most notable works is *La Revolucion Filipina*, the full-length Filipino neo-ethnic ballet which premiered in 1996 at the CCP Main Theater before touring the West Coast of the United States in the same year. The production reaffirmed my understanding of how a Filipino dance company tours overseas and the conditions by which one operates. The 21-day tour with 20 personnel in Los Angeles, Irvine, and San Diego moved from one host family to another, performing to full-house audiences of primarily Filipino-American audiences. All this left a strong mark with me, the multi-tasking tour coordinator and stage manager. My biggest takeaways were: 'What does it actually mean to internationally tour a Filipino production? Who is it for? What defines it as an international performance, and what brings it closer to being world class?'

While I spent only two years at BP, it was enough for me to learn the career path that awaits ballerinas and/or danseurs. I understood that ballet dancers could progress from being a student to a scholar, from a scholar to a junior company member of Ballet Philippines II (BP2), company member to soloist, and then from soloist to principal. After that, one could possibly become resident choreographer, rehearsal master, répétiteur, dance teacher, or even the company's artistic director if they had not retired or switched careers by the age of 35.

Some of the exceptional and international prize-winning BP dancers I worked with 20 years ago, like Georgette Sanchez and Unita Gaye Galiluyo, took a leap to pursue their dance practice as contemporary performers in Europe: Alden Lugnasin's retirement saw his return to his hometown Samar to mentor young regional talents; Dwight Rodrigazo set up a dance school in Bacolod City training contemporary dancers; Christine Crame, one of the first recipients of the company's educational dance programme, became the head of the School of Dance at the College of Saint Benilde. There are also principal dancers and soloists like Cecile Sicangco Ibarolla, Brando Miranda, Camille Ordinario, Gerald Mercado, Katrina Santos-Mercado, Judell de Guzman, and Annette Mariano, who have become certified movement specialists teaching pilates. Their career paths are expected, given the preparations that BP made ahead of their retirements or departures from the company.[5]

Curating dance in Asia and Europe

The idea of curating revealed itself during my time at the Asia-Europe Foundation. Here, I began bringing artists together for short periods of between one and two weeks to exchange ideas and question norms, allowing a space for the status quo. Among my numerous assignments was one where I conceived of a forum called *Pointe to Point*. It involved 20 dancers and 20 new media artists in Tokyo masterminding four intimate pairings of young choreographers and established artists in Baguio City, Kuala Lumpur, Langkawi, Lisbon, and Paris. It culminated with a group of 25 dancers in Warsaw, all of whom were meeting for the first time. I was catapulted into this project after unknowingly preparing for it since my early days at the CCP.

Each step of my journey was determined by my wish to open up a fresh perspective on collaboration, enabling those involved in the process to learn something about themselves and from one another. My objective was to find common ground for diverse individuals who would normally never work together. Such platforms encouraged artists to seek and respond to the impetuses offered by the situation, environment, and people around them.

Upon my move to Berlin as an independent curator and producer, I consumed copious amounts of art through kitchen conversations; conceptual, experimental, avant-garde, classical, and traditional performances in music, opera, dance, and theatre; and openings at the most alternative and most established museums and galleries in the city. I committed to attending festivals in Europe and New York, followed the practices of various artists, and exposed myself to whatever was available for my creative research.

With a vision of encouraging artists to work outside their comfort zones to collectively develop artistic processes that speak from within, I initiated artistic projects from 2007 that included performances, meetings, picnics, and screenings with the likes of RADIALSYSTEM V, Public Art Lab, DMY, Berlin Biennale, Platoon Kunsthalle Berlin, Asia-Pacific Weeks, Bencab Museum, and the Cultural Center of the Philippines, among others.

By 2011, these interventions caught the attention of other professional peers, and I entered the formal museum environment: ArtScience Museum in Singapore and Weltmuseum Wien (World Museum of Vienna). It was in these places that I was able to dig deeper and explore the intersections between performance and the museum. The groundwork I had laid through years of thoroughly observing artistic practices created an intangible repository of research materials founded on human relations.

Work in progress in the museum space

I briefly met Filipina performer Eisa Jocson while working on ZENSORS in Manila in 2009. It was at the residence of the late Carlos Celdran, which was the temporary accommodation I had arranged for my team of Japanese artists. This brief encounter sparked my interest in Eisa's practice as an artist and pole dancer, and I began following her work thereafter.

Years later, Japanese-French artist Shunsuke Francois Nanjo shared with me his 2013 Aurillac exhibition where he presented *Hypermnesia*, an installation looking at memories that was made of a series of metal poles activated by touch-sensitive sensors. He wanted to take the work further by involving a dancer in the work's next iteration. I understood the hint that he wanted to work with Eisa.

Months after this conversation, I received an invitation to curate a monthly programme for ArtScience Museum in Singapore. This would become a series of interdisciplinary art happenings geared towards creating a space for interaction, collaboration, and experimentation.

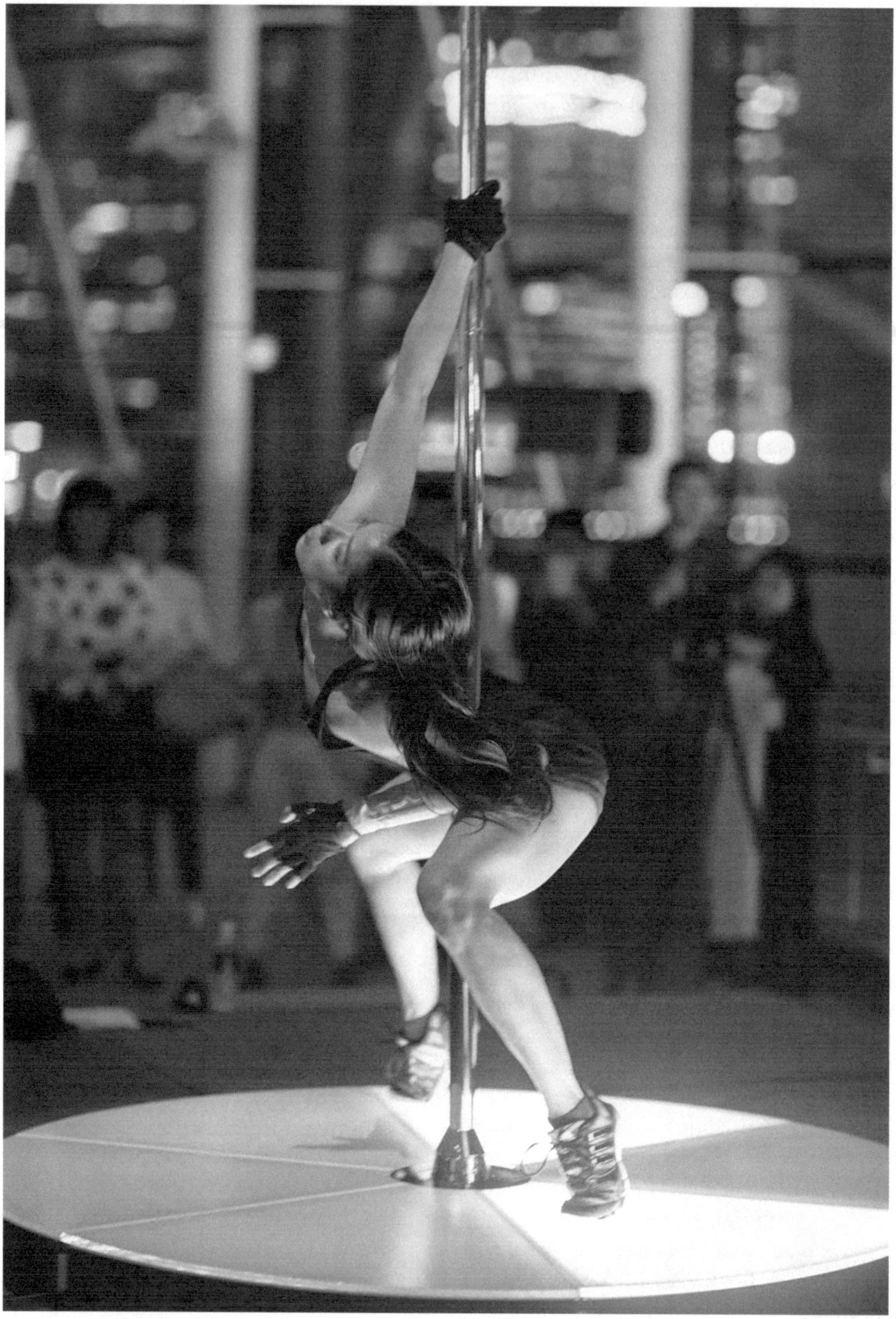

FIGURE 12.1 *Hypermnesia* for ArtScience Late (January 2015)

Source: Curated by Vanini Belarmino, featuring Composit formed by Shunsuke Francois Nanjo, Kensuke Christophe Nanjo, and Nicolas Charbonnier in collaboration with Eisa Jocson

ArtScience Museum at Marina Bay Sands (Singapore)

Photo: Cees Van Toledo

I identified artists who would have the experience and strength and who would welcome new impetuses from people they had yet to work with and to deviate from what was expected of them. Among them was Shunsuke, who formed a team with an aeronautics and civil engineer and a sound designer. I subsequently approached Eisa to complete the composition for *Hypermnesia* Singapore. After a year and half of work, we all met in person in Singapore in February 2015, with three days to set up the interactive pole, get acquainted with one another and the space, and perform two full-house shows.

After this collaborative experiment, I continued following Eisa's work as a cultural journalist, where I witnessed the staging of her work *Host* (2015) for Tanz Im August in Berlin and the Philippine premiere of the piece for Dance.MNL at the CCP. By then, I had been appointed as the general manager of the first Philippine Dance Biennial.

After moving back into the museum space, I invited her to the National Gallery Singapore in my capacity as assistant director (programmes) to respond to an exhibition called 'Between Worlds.' This time, Eisa was commissioned to develop *Luna Tunes*, a 20-minute performance tour in response to the works of 19th-century Filipino master Juan Luna.

The unspoken dance lexicon

Eisa Jocson: archiving, performing, and exhibiting Filipino labour

In the Philippines, ballet plays a significant role in how professional dance practice is understood. If one operates within these conditions and follows the expected career path of a dancer, it is difficult to imagine how taboo subject matters associated with nightclub entertainment – such as pole and macho dancing, Filipino hostesses in Japan, or cover bands – could find a place within the pristine confines of the Philippine dance scene. For someone to position these topics at the centre of their artistic practice appears a remote possibility.

Yet Eisa Jocson has moved beyond what is expected from the best of Philippine pride and away from the beautiful and romanticized images of cultural representation. She has been confronting her audiences with a body of work that 'exposes body politics in the service and entertainment industry as seen through the unique socioeconomic lens of the Philippines'. ("Artist Bio", n.d.) Her trilogy of performances, *Death of a Pole Dancer* (2011), *Macho Dancer* (2013), and *Host* (2015), has garnered immense success, and the pieces are often presented as a double or triple bill by international festivals in Europe, the United States, and Asia ("Seduction of the Economic Body", 2015).

With research focused on and heavily invested in learning about how the body moves and the conditions in which it moves, this young artist is a product of the Philippine High School for the Arts (PHSA)[6] and University of the Philippines, Diliman, College of Fine Arts. She approaches subjects from within movement languages specific to each context. She acknowledges how her ballet training has been formative in her physical and movement orientation and credits her visual arts education and sculpture major for giving her access to theoretical discourse.

Eisa explores modes of presentations in the contexts of performance and contemporary art. Commissions from arts institutions and biennales have enabled her to develop works like *Happyland Series* (2017–present), made up of her renditions of Disney fairy tales. The series consists of *Happy Land: Part 1 Princess* (2017), *Your Highness* (2017), *Becoming White*

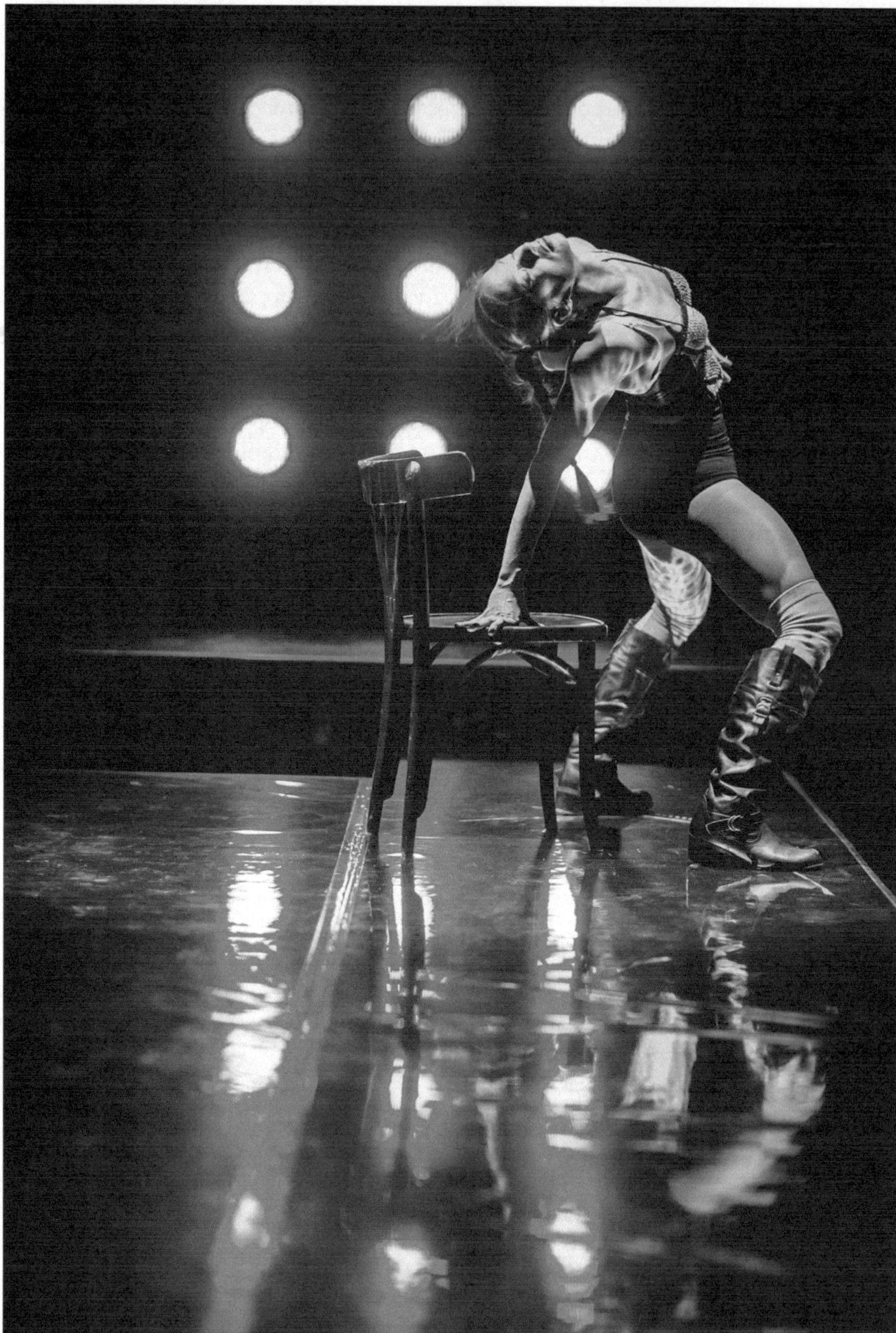

FIGURE 12.2 *Host* (2015)

Source: Eisa Jocson

Tanzhaus-nrw, Dusseldorf

Image courtesy of artist

(2018), and *Manila Zoo* (2020). *The Filipino Superwoman Band* (*2019*), which questions how one might look at or hear the voices of Filipino cover bands across the globe, looks at another aspect of exported Filipino talent. A recipient of the CCP Thirteen Artist Award (2018) and Hugo Boss Asia Art Prize (2019), her investigation of the economic body creates a new, previously unspoken lexicon in dance.

Host (2015) had its Philippine premiere at the CCP Little Theater in June 2016 when Paul Morales, artistic director of BP, was sharing the directorship of Dance.MNL with Ronilo Jaynario and Lisa Macuja-Elizalde, artistic directors of PBT and Ballet Manila, respectively. A rarity in the ballet-oriented Philippine stage, Paul's decision to offer an expanded range of Filipino dance was a warm welcome to a generation of fresh thinkers like Eisa. In the words of the prize-winning artist-choreographer:

> I think to put that movement language at the CCP Little Theater Stage is an important statement. It also confronts what kinds of vocabularies and bodies can perform in such a stage. Predominantly ballet has the privilege to be placed in the centre. While these forms that are currently being utilized economically by Filipino/Filipina migrant workers are deemed less valuable, or not culturally significant.
>
> *(E. Jocson, personal communication, May 17, 2020)*

Unlike choreographers bound by the tradition of ballet, Eisa's practice is informed by the transference between dance and visual culture. She actively departs from the norm, going beyond narrative works to define her own artistic trajectory. Her independence goes beyond being a freelance artist: she is a free thinker who defines her own artistic language while earning her livelihood from it. While she is adamant about keeping her base in the Philippines and employing particularities drawn from the Filipino body and its labour, her works continue to be presented abroad – her concerns centre on how the body moves, the conditions that make it move, and social mobility and non-mobility, as well as movement out of Philippines in the form of migrant work. In her words:

> I am interested in current body languages formed by social conditions. It follows that our body politics is a reflection of the society we live. What is shared between pole, macho, host is that capital is a driving force of movement in the body and in into spatial geographies. I work between Manila and where there is support for what I do; Belgium, Germany, Switzerland, Japan, Hong Kong etc.
>
> *("Seduction of the Economic Body", 2015)*

Her performances are focused on the investigation of the body, and her intent is to challenge assumptions and perceptions of a body's value. This is determined by where it is placed: its location and the conditions it operates under. The manifestations of her work, as well as their presentations in the theatre and museum, are symbolic representations of relocating value – including the perceived values of dance, performance, and art. They ask one to consider what Filipino dance could be, encouraging us to possibly expand our understanding of the so-called 'cultured'.

Eisa also confirms that she does not concern herself with being liked or loved by her audience. The reason that she puts forth representations of the likes of pole dancing, macho dancing, and hostesses in the theatre is because theatre is where certain boundaries can be broken. It provides space for worldviews that one embodies. In her own words:

> What I'm trying to propose is that the human body is actually quite fragile depending on the context that it is in. It also has the strength to change into what it wants to be, that we are always evolving. I'm interested on how we move from within our bodies, into outside, into transnational. Outward social mobility, that's why I focus on labour. There's this certain type of hospitability, Body as a host. It's about this, as opposed to the aesthetics of the form.[7]

Away from the formalist system whereby a professional dancer is mentored by certified classical dance teachers, Eisa credits the real-life players of the bodies she's chosen to archive in her system as her teachers. She explains:

> There is a long and intimate transmission process between my teachers working in the field (macho clubs, host club in Tokyo) that results in my embodiment of their dance practices. Dancing their dances does not equal to telling their stories. These marginal movement practices loaded by their specific context become tools in destabilizing dominant and comfortable ways of seeing. All three works operate in the dissonance between perceived notions of reality and performed embodiment.
>
> *(E. Jocson, personal communication, May 17, 2020)*

Eisa is, therefore, an exception: she is one body, a singular dancer/choreographer, an outlier, who took a giant leap from the first position to articulate the multilingual, multi-tasking Filipino body through dance.

The making of the trilogy: *Death of a Pole Dancer, Macho Dancer* and *Host*

Eisa was part of the first batch of students at a pole dance studio at a time when the sport was still heavily stigmatized. She became a pole dance instructor after two years and then saw the potential of pole dancing as a tool for her artistic practice. In her words:

> Pole dancing being heavily charged with its own history was in an interesting period of transition and appropriation from the nightclubs to the fitness industry. The shift lies in the embodied engagement with pole dancing. The physical labor, monetary investment, change of lifestyle to become a 'pole dancer' is exposed. For those practicing pole dancing in this context, it is no longer an image outside of their sphere but a lived practice that brings with it a different set of conditions as well as community formation.

Eisa created her first solo performance work, *Death of the Pole Dancer* (2011), through the invitation of Singaporean dramaturg Tang Fu Kuen for the *In Transit* performing arts festival at the Haus der Kulturen der Welt, Berlin. The work interrogated how one views what is on show – the performance negotiated notions such as voyeurism and restraint, vulnerability and violence, sexuality and power, bringing audiences to reflect on what they witness: a woman in the act of pole dancing.

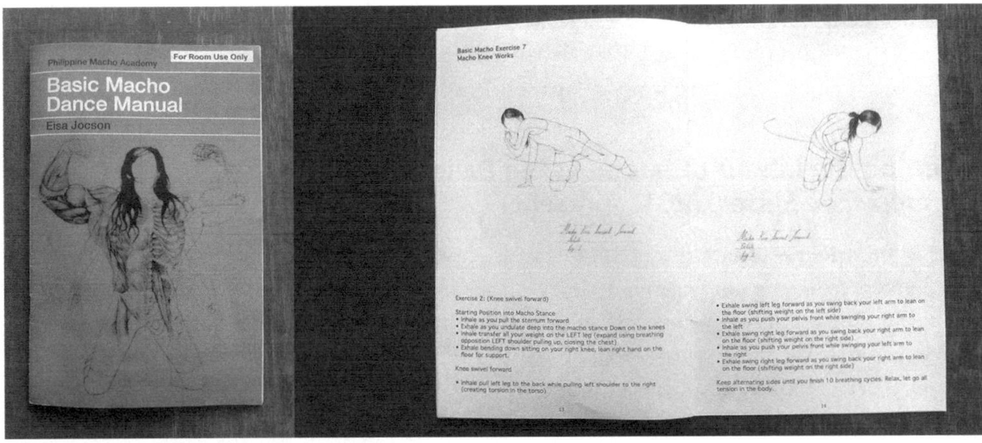

FIGURE 12.3 *Philippine Macho Academy* (2014)

Source: Eisa Jocson

Vargas Museum, University of the Philippines, Diliman, Quezon City

Image courtesy of artist

In 2012, she moved away from pole dance and learned what can be considered oppositely gendered in the realm of night work – macho dancing. She visited several macho clubs regularly, inviting good macho dancers to train her. This year of training transformed the movement practice of her body. She said, 'I was stepping outside my given body politics as a woman as well as expanding spatial sphere into the macho nightclubs of Manila' (E. Jocson, personal communication, May 17, 2020). Made between Manila and Brussels, Eisa's *Macho Dancer* (2013) premiered in Brussels and toured Europe, America, and Asia. It also won the prestigious Zurcher Kantonalbank Acknowledgement Prize at the Zurich Theater Spektakel in 2013.

Beyond performance, the choreographer also approached Patrick Flores, then curator of the Vargas Museum, to make an exhibition about the process behind *Macho Dancer*. This resulted in a solo exhibition in March 2014 titled *PMA; Philippine Macho Academy*, which proposed a fictional institution of masculinity.

While continuing to tour the two works, Eisa became curious about the existing relations of Philippines and Japan. During a residency with the Saison Foundation in November 2014, she met a transgender Filipina hostess who taught her one of her dances. At the same time, Eisa was learning *Nihon Buyo*[8] as a way to understand Japan through the body. Fascinated with how Filipino entertainers negotiated their identities to fit the demands of Japanese clients, Eisa created *Host* (2015) in Tokyo, Brussels, Yokohama, and Manila before it premiered in Dusseldorf. To create the piece, Eisa 'had to reorient my body away from macho dancing into a malleable vessel that is versed on transcultural notions of feminine representations' (E. Jocson, personal communication, May 17, 2020).

The yearlong process of preparation behind each work provided Eisa the time to learn and digest each vocabulary of movement. The longer she used it, the easier it was for her to recall the movements. *Host* was particularly challenging to learn: the hyper-feminine vocabularies of hostesses were located in multiple geographies. The Japayukis, or Filipino entertainers, had to cater to the Japanese salaryman. There were two cultures and contexts: for entertainment, there was the Sexbomb Dancer, whose urban Philippine femininity was exported in this circuit of hostess clubs, and there was the Japanese context, to which these women had to adjust.

Returning the labour of happiness to Ballet Philippines, the *Happyland* Series, (2017–present)

In her 2008 undergraduate thesis, Eisa examined the psychological health of BP when half of its dancers from the core company left for jobs in Hong Kong Disneyland. The premiere Philippine dance company was left with apprentices. Hong Kong Disneyland recruited a great number from the Filipino artistic community, and the loss of talents cut across the fields of theatre, dance, and even music.

Developed over the ten years since this exodus from BP, the *Happyland* Series is an exploration of what Eisa refers to as the 'labour of happiness', something that the Filipino artists who work at Disneyland have to repetitively perform. Disneyland engages Filipino artists for supporting roles that surround beautiful white princesses, and Eisa, like in her previous works, spotlights the unnoticed to ask audiences to look beyond 'happily ever after'. The series, which manifests both in theatre and gallery spaces, examines the conditions and formations that are unknowingly put in place during childhood through the dominance of Western fairytales. By raising the issue of racial profiling, it points out the

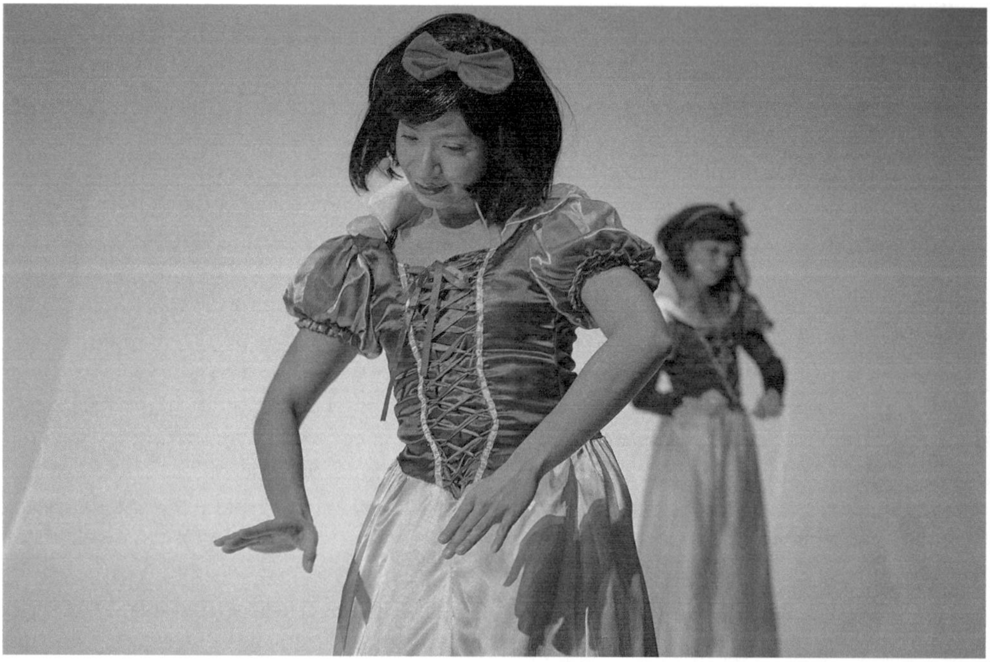

FIGURE 12.4 *Princess Studies* (2017)

Source: Eisa Jocson

Frankfurter Positionen 2017, Mousonturm, Frankfurt

Image courtesy of artist

disparate narratives that shape working conditions and the ideas of labour and value. The series has since seen the realization of *Happy Land: Part 1: Princess* (2017), part 2: *Your Highness* (2017), *Becoming White* (2018), and part 3: *Manila Zoo* (2020).

For *Your Highness*, Eisa brought together an unlikely ensemble of five young ballet dancers from BP: Gia Gequinto, Stephanie Cabral, Maila Habagat, Alexis Piel, and Carlo Padoga. At that point in time, this new generation of dancers were aware of the new reality for professional dancers: that at some point, they, too, would follow the footsteps of those who came before them to go to Disneyland or join a cruise ship. Such a trajectory would have been unthinkable to the previous generation of dancers from 20 years ago, when overseas labour employment was not part of the continuity of dance practice.

The work was realized with the help of Paul Morales, BP's former artistic director, who provided a space for a different way of working – a more experimental and reflective approach that could accommodate a different choreographic practice that involved the dancers talking and articulating their opinions. For dancers accustomed to taking ballet classes day in and day out, this came initially as a shock. They had to reorient themselves to a process that asked of them to think about their conditions beyond just being present for class. Eisa explains:

> the process of developing the skill of reflection posed a huge challenge for ballet dancers who were never trained to use their voices for.
>
> While *Your Highness* was not overtly criticizing ballet, the work manifests criticality by revealing certain things inside the fact that most of the vocabulary used in teaching is in French. It's part of a daily practice. It becomes a part of who you are. In the work, dancing ballet while naming the French terminology of each step becomes the material of the performance making audible coloniality. I am interested in showing this disruption between the form, the spoken vocabulary, and the Filipino body.
>
> My thinking then is that it doesn't matter if they don't fully comprehend the piece when it was made. What matters is the process that we went through together. It was a challenge to question the institution of ballet, which was at the core of their identity during that time. It was not an easy process, but they were willing and open to it.
>
> *(E. Jocson, personal communication, May 17, 2020)*

The work was geared towards hijacking: corrupting what one knows; acknowledging colonial heritage; and using it to be reflective, grounded, and proactive, as Eisa brought the very labour of happiness back home to where it was originally harnessed.

The blackbox and white cube

As Eisa develops the *Happyland Series* and expands her vocabulary, her works likewise begin to occupy different spaces of discourse. She continues to navigate the space between the theatre and museum, and the overlaps of performance and the visual become more present, the meeting between the ephemeral and material. This is also manifested in *Filipina Superwoman Band*, a work that Eisa created for Sharjah Biennale in 2019 where the presentation consists of both performative and exhibitionary formats. Sculptures, video, drawings, objects, and performance installations offer an extension to her body and body of work.

Eisa recognizes the importance of reaching a broad spectrum of audiences. She observes that within the visual arts, audiences can go in and out of durational performances and that there is no obligation for them to stay throughout. They have agency over how to experience the work. As the seasoned multi-disciplinary artist puts it:

> The theatre offers the ephemeral and collective experience as opposed to presenting a collection of things experienced individually. . . . The aspect of being looked at and looking is confined with a given time frame. As soon as the audience step out of the theater the experience is the only thing that remains. Even if there is a video documentation, it stays as an archive or a shadow of the work.
>
> The visual arts and the theatre space have different social contracts, traditions and space-time parameters that one has to consider in creating work. Durationality, for example, is usually operative in the contemporary performance-visual artist context rather than a western theater context. When you propose a durational performance within the theatre, you're also expecting the audience to somehow accompany you throughout which is not the case in visual arts. . . . Of course one can depart from the standard or normal.
>
> *(E. Jocson, personal communication, May 17, 2020)*

Nothing is standard or normal with this extraordinary artist, who decided on setting up her own language foundation. As Eisa Jocson enters a decade of steady international success both in dance and contemporary art, I would like to consider the words of Dr. Lucresia Kasilag, the first CCP President and National Artist for Music, at a commencement address for PHSA, Eisa's alma mater, in 1982, years before Eisa was born or entered the art world. Her address can be seen as both a foreshadowing of events and a warning that will conclude this chapter: 'As young artists who shall soon be the leaders in your own chosen fields of endeavor in the days to come, you cannot afford to fail, or falter, or condescend to lesser motives other than being true and honest to your art in the pursuit of artistic excellence' (De la Torre, 1984, p. 157).

Notes

1 The Cultural Center of the Philippines (Sentrong Pangkultura ng Pilipinas, CCP) is a national institution that aims to preserve, develop, and promote Philippine culture. Founded by former First Lady Imelda Romualdez Marcos in 1969, it upholds the Filipino practice in the seven arts of music, dance, theatre, literature, visual arts, film, and architecture alongside the presentation of international productions. It houses what many consider the best in Filipino arts while allowing for organic interdisciplinary exchanges.

2 The CCP has a total of nine resident companies. Apart from the four dance companies noted previously, there are the four music groups – Philippine Madrigal Singers, the UST Symphony Orchestra, National Music Competition for Young Artists Foundation (NAMCYA), and Philippine Philharmonic Orchestra (PPO) – and a theatre company, Tanghalang Pilipino (TP). Among all these groups, BP, PPO, and TP are housed inside the centre.

3 Ballet Philippines was founded in 1969 by Philippine National Artist for Dance Alice Reyes and Eddie Elejar the same year the CCP was inaugurated. It is the first resident dance company of the CCP and the first to professionalize dance by providing full-time employment to its company of dancers and staff. It boasts more than 500 original works that range from full-length classical ballets to Filipino masterwork.

4 The Order of National Artist of the Philippines is the highest award conferred to recognize a person's contribution to the arts. The categories are for the seven arts of music, dance, theatre, literature, visual

arts, film, and architecture. It is administered by the CCP and the National Commission for Culture and the Arts.

5 Ballet Philippines management, under the leadership of Jardin and Cecile Manikan in the mid-90s, had a vision of preparing its company members for alternative careers after retirement. For instance, Cecile Sicangco-Ibarolla was sent to New York City in the summer of 1995 to attend pilates training. This eventually paved the way for several former ballerinas to pursue a career still connected to mastery of the body. Another milestone was the establishment of the dance degree programme at the College of Saint Benilde in 1996, which enabled them to study on full scholarship and work at the same time.

6 The Philippine High School for the Arts (PHSA) is a special secondary art boarding school established at the National Arts Center in Los Banos, Laguna, in 1977. The school's special curriculum is geared to the early recognition and development of highly talented children exceptionally gifted in the arts, thus providing a continuing source of artists of excellence and leaders in the preservation and promotion of the Filipino artistic and cultural heritage. (Section 2, Executive Order No. 420), www.phsa.edu.ph/about/mandate.html.

7 On 22 June 2016, Eisa Jocson joined a panel discussion for Dance.MNL's Conference at the Cultural Center of the Philippines, where she presented *The Economic Body*. At this panel, she addressed a query from a dance student who inquired on how Eisa made people fall in love with her art and how she managed to change perspectives on pole and macho dancing.

8 *Nihon Buyo* is a Japanese dance that belongs to the classical Japanese performing arts. Eisa incorporated it in her choreographic work *Host*, revealing a feminine Japanese character who moved slowly in her hot pink kimono, with a fan that functioned as an extension of her hands and arms.

References

Artist Bio. (n.d.). Retrieved May 2020, from Eisa Jocson: https://eisajocson.wordpress.com/about-2/

De la Torre, V. (1984). *Cultural Center of the Philippines: Crystal Years*. Manila, Philippines: Cultural Center of the Philippines.

gratz.com/pages/featured-studio-series-december-2011-integrated-body-arts-pilates

Jocson, E. (2020, May 17). Skype interview.

Seduction of the Economic Body: Vanini Belarmino in Conversation with Eisa Jocson. (2015). Retrieved May 2020, from Tanzconnexions: Dance in Asia Pacific: www.goethe.de/ins/id/lp/prj/tco/arc/por/en14946026.htm

Singapore

13

DANCESCAPISM

On dance, young people, and choice in the Lion City

Peter Gn

Worldwide, there is increasing reference to dance as a meaningful and creative form of experience for youth. Dance allows them to develop skills, knowledge, and understanding as critically reflective dance performers, dance-makers, viewers, and investigators. Besides offering the joy of self-expression, dance provides new forms of learning and facilitates the development of problem-solving skills which access multiple intelligences. It can be a powerful medium through which young people come to terms with themselves and others.

In this chapter, 'dance education' refers to dance taught as an art form, whether in school-based programmes or as a co-curricular activity (CCA) to effect learning in and about dance, embracing youth participation and all aspects of dance that have educational value. I will look at the current landscape, the place dance has in youths' lives, and the value of the various platforms that support it. The underscoring tenet is that if dance is to function as a vital experience for youth, shared responsibility should occur in collaboration with educators, dance professionals, and advocates. The scope of this chapter and the examples cited are by no means exhaustive.

Dance in general education

Dance education in Singapore has for many years had a traditional focus on movement, technique, and virtuosity. The number of dance organizations and professional dancers has risen, with arts institutions like the LASALLE College of the Arts, the Nanyang Academy of Fine Arts, and the School of the Arts Singapore significantly contributing to the country's growing number of young dancers.

The purpose of dance in general education is not to train professional dancers. Dance, like visual art, music, and sport, has an important role in the holistic education provided in Singapore schools. Yet despite the many voices in support of dance education, in reality it remains in an arguably marginal position. For many years, dance has not been available as an academic subject within the general education system and has remained either a CCA or part of physical education (PE). Public impression of dance is also mixed, with attitudes ranging from the negative and dismissive to those that deem dance something

DOI: 10.4324/9781003160007-23

only for the talented. Career prospects related to dance are also perceived by many parents as discouraging.

Currently, arts subjects offered in the mainstream school curriculum are music, visual arts, and drama. Dance is available only as a CCA and not an academic subject. Many forms of dance CCAs exist in Singapore schools, including Chinese, Malay, Indian, ballet, hip-hop, ballroom, and more. Dance CCAs aim to develop a variety of generic skills like collaboration, communication, critical thinking, and creativity as well as aesthetic appreciation. These, along with the nurturing of positive values and attitudes in students, are seen as a foundation for lifelong learning. Advocacy has an important role to ensure that the value of dance as a meaningful, educational experience is recognized. Yet currently, though dance is recognized as integral to the customs and traditions of the matrix of Singapore cultures, it is not widely understood, supported, or valued across all schools. Misconceptions exist regarding the content and outcomes for students of dance CCAs; for example, the content of dance CCAs across schools tends to be rather disparate; with mostly a focus on the acquisition of particular dance skills and techniques or learning choreography that is mostly steered towards participation in performance platforms, among them the Singapore Youth Festival (SYF) and other commercial (or non-commercial) dance competitions.

Dance as a CCA across the four tiers of the education system in Singapore schools (primary, secondary, junior college, pre-university and university) currently takes place with resources, pedagogy, and student demographics varying significantly. Beyond the dance CCA context, schools do also include dance in their various student programmes in diverse ways, having full autonomy to do so. As dance CCA programmes tend to focus on stage performances and competitions as goals, dance teaching by the freelance instructors hired by schools often proceeds along one-dimensional, content-driven lines, leaning heavily towards skill acquisition and taught choreography. A sizeable number of dance instructors of CCA programmes are young graduates from arts institutions like the Nanyang Academy of Fine Arts (NAFA) and LASALLE College of the Arts. Some are independent artists, while others are from commercial dance studios. Dance within the schools' CCA framework thus may be said to provide only short-term development such that even if students derive enjoyment and satisfaction, this is not necessarily always process-driven or student-centric. And sustainability is usually contingent on the individual school's support and factors related to funding or whether there are adequate numbers of interested students to make the dance CCA viable.

In primary schools, dance can sometimes be part of an integrated platform occurring within a range of subject areas. Children may create some movement associated with an art, music, or even science class, but as the primary school curriculum does not mandate the incorporation of dance or movement across all subjects, this depends largely on the teacher and the classroom topics covered. A small dance component is incorporated in PE programmes, as well as the Programme for Active Learning (PAL). PAL is part of the curriculum of all primary schools, and through it, students are exposed to varied and fun learning experiences in four domains: sports and games, outdoor education, performing arts, and visual arts. Students also acquire socio-emotional competencies like respecting others and responsible decision-making. PAL also nurtures confidence, curiosity, and cooperation skills in students. There are dance CCAs in many primary schools, and these offer opportunities for children to learn and perform different genres of dance and represent their schools at various events.

Many secondary schools offer a range of dance CCAs from the cultural to contemporary. Dance in the secondary schools' PE curriculum is optional. A number of mainstream secondary schools include dance education as enrichment in their regular curricula but not as an academic subject. Again, most conduct traditional skill training in dance CCA classes, taught mainly by freelance dance instructors. A large majority focus on the preparation and presentation of dance works for the stage. Junior colleges presently have non-compulsory dance CCAs, and they participate in a host of different dance activities and performances, whether school-based or otherwise. There are no O- or A-Level dance courses offered in Singapore.

More schools are now staging dance productions in public venues and theatres, and some have been sending their dance CCA groups overseas annually to participate in dance competitions and exchange programmes. Some schools participate in mass dance displays (featuring hundreds of students) at the National Day Parade. Schools are also selected to showcase their dance performances at various platforms outside the school, whether related to the Ministry of Education (MOE) or otherwise. Individual schools may organize youth dance camps and workshops (at the school or cluster levels) where the different groups of students can mix and learn each other's dances in a cooperative environment. In addition to these programmes, schools also participate in the National Arts Council's Arts Education Programme (NAC-AEP).

NAC Arts Education Programme

The AEP is a collective of specially selected arts education programmes by professional artists and arts groups that schools can purchase using an annual grant to subsidize part of the cost. The NAC-AEP covers six different art forms (dance, music, theatre, film and multimedia, visual arts, and literary arts) and consists of three categories: Arts Exposure, Arts Experience, and Arts Excursion. The NAC-AEP gives students access to quality arts education that enables them to develop creativity and expression through the arts, as well as deepening their understanding of arts and culture.

For dance, under Arts Exposure, young people are introduced via 30- to 40-minute customized performances by professional dance companies on the NAC-AEP list. Each performance is followed by a talk on the background and development of the segments presented. Arts Experience gives smaller groups of students a hands-on experience in dance in, and through, an interactive environment. The Arts Excursion programmes enable students to attend dance performances held at arts venues or visit museums, galleries, art centres, artists' studios, and theatres. Feed Your Imagination (FYI) under the Arts Excursion Scheme is a partnership with Esplanade Theatres on the Bay, for example, and it offers young people the chance to view quality productions with strong educational content by professional dance groups. Other initiatives in schools such as the NAC's Artist Mentor Scheme support collaboration between artists and schools in developing arts education projects.

Teachers also bring students to participate in other dance events that provide a special experience. The interaction of thinking and feeling between the performers and the audience at a live performance is invaluable and develops critical appreciation of a dance. While students should not learn about dance by simply watching performances, this is certainly a recommended supplement for CCA students where they can meet the dancers, choreographers, and costume and set designers, amongst others.

Beyond the school

Outside of mainstream schools, there is a burgeoning interest in dance among young people, as evidenced by the steady rise in participation rates in all genres increasingly featured prominently in various dance platforms and events. Many commercial dance competitions have also emerged, with attractive incentives in cash and exposure. These competitions frequently feature street dance forms popular with the youth and, with their innovative approaches, inspire a huge following. The Super 24, a dance competition for young people introduced in 2013 as a new and innovative dance platform by O School, a social enterprise, showcases not just high standards of dancing but also the depth of teamwork and unity. Each of the Super 24 teams' crews from across the country consists of exactly 24 members dancing within an 8-by-8-metre square area, performing a 90-second routine. To add to these challenges, Super 24 replaces the conventional practice of 'front view' judging with four-sided judging, where movement sequences will be noticed. Hundreds of groups from schools take part in the competition, and the finals are often held at high-profile arena-type venues that draw large crowds.

Dance festivals

Dance festivals often entail active collaboration on all levels (whether between individuals, schools, venues, and organizations) to promote the diversity and understanding of ideas, opinions, and attitudes. Sometimes they aim to place young people at the heart – so besides being a showcase for young people's creativity and expression, they also aim to give them key responsibility from the promotion of the festival to performance. Consider the highly successful da:ns festival, an initiative by the Esplanade Theatres on the Bay and annual event since 2006. In the festival's efforts to increase young people's awareness of, and access to, dance, the festival forges new partnerships between those providing and supporting dance activities for young people through working across the arts, community, education, sport, and youth sectors. The festival celebrates the spirit of movement and seeks to unite dance lovers from everywhere in Singapore and the world. In particular, through connecting young people with and through dance, the festival allows them to participate in a wide range of dance workshops or dance-related activities, whether for fun, skills acquisition, development of an understanding in dance, or the display of their work in one of Singapore's eminent performing arts venues. Footwork, a platform under the da:ns festival umbrella, for instance, encourages young people to discover the joy of movement. Juxtaposing showcases by professional dance companies both from Singapore and abroad, it offers workshop opportunities for Singaporeans in a range of introductory classes and dance styles from street-dance to *bhangra* and belly dancing catering to all ages.

CONTACT Contemporary Dance Festival has become another annual event initiated in 2010 by The Human Expression (T.H.E) Dance Company. Usually held in late June, CONTACT features a wide range of dance activities aimed at promoting the discovery, learning, and exchange of dance amongst people and enthusiasts. At this festival, young people get to immerse themselves in many technique classes, workshops, and performances conducted both by the company's artists as well as those from the international arena.

The annual Singapore Youth Festival, an MOE flagship for nurturing lifelong interest in the arts, has been seen as playing a significant role in the aesthetics education offered in Singapore schools. The SYF is the largest annual school event in Singapore organized by the MOE to celebrate the achievements of youth in arts CCAs. It targets youth from primary and secondary schools to junior colleges and offers performances as well as opportunities through showcases of music, art, drama, and dance talents from the schools. Annually, thousands of young people participate in the SYF performances and activities.

Dance by young people has always been showcased prominently at the SYF. Starting in April and culminating in a celebration phase in July, students in schools across Singapore who are involved in arts CCAs are engaged in either a benchmarking exercise in April (the Arts Presentation) or a performance showcase in July. The many-faceted dance offerings at the SYF include colourful multi-ethnic costumes and myriad performances involving multiple dance genres. Over the past four decades, dance at the SYF has grown in scope, diversity, and quality of student achievements. There is currently no platform for young student choreographers at the SYF, and professional dance choreographers and instructors are hired to stage large-scale works. Given the mammoth participation base and the emphasis on excellence and perfection, students have been engaged mainly as audience and performers.

University and polytechnic student dance groups

At the tertiary level, dance performance opportunities for young people abound. The National University of Singapore (NUS) Arts Festival started in 1998 as a showcase of student talents. As part of the centenary celebrations of the University in 2006, the Festival was relaunched and came to incorporate faculty and practising local and international artists alongside student performers. At present, the Festival draws artists and collaborators from across campus and around the world and offers the chance for students to both engage in and appreciate dance. The dance programme at the Festival typically features ticketed public performances mainly in the University Cultural Centre Hall and Theatre. Performances range from the cultural to contemporary and street dance, and student choreography is often encouraged. The dance groups at the Festival are usually drawn from the student body; thus, they comprise young people from myriad dance CCA programmes within the varsity – key among them are the NUS Dance Ensemble, NUS Dance Synergy, NUS Chinese Dance, NUS Ilsa Tari, NUS Indian Dance, and NUS Dance Blast!, all of which are managed by the NUS Centre for the Arts.

Seeking to be inclusive and fostering a welcoming environment that allows new or inexperienced young dancers to grow and learn, each NUS dance group has distinctive features. Established in 1977, NUS Chinese Dance has a repertoire ranging from the technically demanding traditional to contemporary Chinese dance. Formed in 2001 and with dance moves that are sleek, sharp, and strong yet smooth, NUS Dance Blast!'s repertoire includes hip-hop dance and funk. The NUS Dance Ensemble was formed in 1991 and remains a platform for young dancers and choreographers. It incorporates styles as diverse as contemporary, jazz, and ballet in its unconventional repertoire. Ethnic infusions in its choreography are a signature trademark. NUS Dance Synergy promotes dance as a modern art form that incorporates elements of jazz, lyrical, and contemporary dance. Established in the early 80s, NUS Malay Dance was renamed NUS Ilsa Tari in 1998. Today it performs traditional dances such as *inang*, *ronggeng*, *zapin*, *lambak*, and *silat*. It also incorporates contemporary dance elements in its repertoire, with choreography carrying a blend of the traditional and contemporary. NUS Indian Dance was established in 1977 and has prided itself in fostering an environment that allows young people to learn the Indian culture through dance.

Organized by Nanyang Technological University's (NTU) Cultural Activities Club (CAC), the NTU Arts Festival (formerly Nanyang Arts Festival) made its debut in 2003 as a special project for promoting arts expression and appreciation. The Festival typically involves the 23 cultural and arts groups under the CCA. At NTU, young people can join dance interest groups that offer everything from hip-hop, breakdance, and contemporary dance to Chinese dance, dancesport, salsa, and funk. Now in its sixth year, the Singapore Management University (SMU) Arts Festival annually showcases creative expression through a range of art forms like music, dance, theatre, and literary as well as visual arts, bringing the arts to the public audience. The dance clubs at the university include SMU Ardiente that focuses on Latin dance; SMU Funk Movement, a street-style dance club that specializes in popping and locking; and SMU INDANCITY contemporary dance club.

The Republic Polytechnic's week-long dance festival, Momentum, is held annually in May, aiming to encourage creativity in dance and promote the original works of the dance groups from not only the polytechnic but also mainstream schools as well as institutions of higher learning. The Festival gives these young people the platform to explore and showcase genres like ballet, breakdancing, and cultural and contemporary dance. The Momentum Dance festival also features workshops and exhibitions.

Dance is also thriving in the Lion City's polytechnics, and examples are mentioned here. The Republic Cultural Centre (TRCC) is an integral part of Republic Polytechnic. The Centre presents a spectrum of programmes to a diverse audience. It sees itself as the focal point for the artistic development of the polytechnic's students and as a cultural space that integrates the community through arts appreciation. The palette of dance groups includes the Style Groovaz Crew, Latin Dancesport group, Modern Dance, Salsa, Dharma Endari, and Indian Cultural Dance, all of which offer the young people a chance to shine in the genre they are keen on or talented in. Elsewhere, Ngee Ann Polytechnic also offers young people a good spread of dance CCAs such as contemporary dance, breakdancing, Chinese dance, dancesport, and hip-hop.

Emerging choreographers

The platforms available for emerging young choreographers are gradually gaining visibility and presenting work at such platforms as Dance Nucleus, M1 Contact Contemporary Dance Festival, Got to Move, M1 Fringe Festival, and Singapore Chinese Dance Theatre's 'Emergence'.

Pre-professional dancers

The Singapore Dance Theatre's (SDT) Scholars Programme is a pre-professional training programme created especially for young dancers between the ages 13 and 19 who are considering a professional dance career. Acceptance to this pre-professional programme is through invitation or audition only, and the programme guides young dancers in pursuing their dance goals while grooming the next generation of professional Singaporean dancers. Guided and taught by the artistic director and selected faculty members, young dancers receive training in ballet technique, including pointe work, variations, and contemporary dance, building them up to become well-rounded and versatile dancers. Also, under the SDT is the Ballet Associates Course catering to young people who intend to enhance and develop their dance skills and techniques. The SDT also offers programmes for schools of all levels, allowing young people to peer into a professional dancers' workplace through offering structured tours of its shoe room and costume sets room and glimpses of the professional dancer's daily routine and working environment.

Founded in 2006 by dance artists Kavitha Krishnan and Juraimy Abu Bakar and its general manager, Imran Manaff, Maya Dance Theatre is a dance company that also conducts classes for young people in the community and offers choreographic opportunities.

Frontier Danceland with artistic director and choreographer Low Mei Yoke is another professional dance company with a track record of reaching out to young people. Contemporary dance workshops are offered, and the PULSE Programme and Dancer's Locker choreographic showcase offer opportunities for youth seeking to advance their technique and gain greater exposure to the difference aspects of contemporary dance and dance production. The company also provides a range of classes for dance enthusiasts, from beginner to professional level. Customized programmes are offered for schools and take the form of workshops, camps, and CCA training in mainstream schools.

T.H.E Dance Company, founded in September 2008 by Artistic Director Kuik Swee Boon, offers workshops to schools and young people, focusing on contemporary dance, ballet, and body conditioning. As part of the company's outreach efforts and believing that dance as an art form is an integral part of any community, it organizes outreach performances of excerpts of their repertoire that seek to bring contemporary dance closer to the community. The company hopes that these projects will raise youth awareness and interest in dance. Young dance artists in the T.H.E Second Company are between 16 and 29 and are exposed to performances, festivals, seminars, and other dance-related events. They are guided by members of the main company as active contributors to the development of the local dance scene.

Each of these has provided outstanding dance experiences for young people in Singapore.

Cultural dance

In the world of cultural dance, young people have no fewer opportunities, too. Sri Warisan Som Said Performing Arts, a performing arts company formed in 1997 by renowned Cultural Medallion recipient Som Said, offers music and dance training opportunities for its more than 200 student members. Bhaskar's Arts Academy, a leader in Indian performing arts in Singapore under the artistic direction of Santha Bhaskar, provides a ready platform for talented young artistes and believes in developing the potential of young people in its teaching wing, Nrityalaya Aesthetics Society. The latter has more than 1,000 students of dance, music, and yoga.

ERA Dance Theatre began in 2009 as a company under Osman Abdul Hamid and Azrin Abdul Rahim. The company's cultural performances in major venues such as the Esplanade and Singapore Conference Hall as well as Victoria Theatre feature youth talents, and its new platforms continue to pave the way for youth with potential in Malay dance to hone and showcase their craft. Founded in 1989 by Artistic Director Madam Lim Moi Kim, the Singapore Chinese Dance Theatre (SCDT), formerly the Singapore Hokkien Huay Kuan Dance Theatre, began as a vocational troupe offering Chinese dance courses and performing opportunities for dance enthusiasts under the Singapore Hokkien Huay Kuan Arts and Cultural Troupe. Currently it provides opportunities for young people to learn Chinese classical dance and Chinese ethnic folk dance. Also providing scores of classes in Chinese dance training for young people is the Dance Ensemble Singapore Arts (DES Arts) Company, well-known in Singapore as a leading contemporary Chinese dance company under the guidance of its founder, Mdm Yan Choong Lian.

The Singapore Ballet Academy (SBA), established in 1958 by Frances Poh, Vernon Martinus, and Soo Nee Goh, offers ballet programmes for beginners of ages 4–5 years to the vocational graded level. Students are trained in the Royal Academy of Dance (RAD) syllabi. As students progress, the training focuses on developing the strength, flexibility, endurance, discipline, and body control demanded by ballet, modern, and contemporary dance styles. Talented students with potential to take up ballet seriously will receive close guidance and intensive training to prepare them for a professional dancing or teaching career.

O School, a performing arts centre, is a social enterprise set up in 2006 and supported by the Ministry of Community, Youth and Sports ComCare Enterprise Fund. O School seeks to generate funds for low-income Singapore youth to complete their secondary school education via City College, providing training and employment opportunities for talented youth. At O School, children and young people have access to a range of street-dance classes.

Perceived challenges

Though youth dance is flourishing in Singapore, with many commentators lauding its significant breadth and diversity and the superlative works by youth dance groups, there are unaddressed concerns and important realities that could be obscured by rosy rhetoric.

Communication between groups remains limited, and collective information on the variety of activity across Singapore appears to be fragmented, sometimes with few opportunities for youth dance groups across the different institutions to come together to collaborate and perform their joint work. Divisions exist between youth dance taking place with the support of dance companies in the community and schools. Other perceived challenges include a shortage of support and opportunities for development of choreography skills among interested young people, presentation and production skills, and work with special needs groups. Regular support structures and opportunities are needed in the areas of networking, information, career advice, and coordinated showcases, for instance, with younger-generation dancers taking their work to a variety of audiences and performing alongside professional dance groups. For some, these opportunities exist, but for others, they are limited, leaving youth dance leaders, coordinators, and instructors often feeling the urgency for increased professional development and support.

Multiple dance strategies can only be beneficial when implemented in a sustained manner and as a planned programme. At the core, more than just harnessing youth energy in the arts and seeking to draw out the best in them, engagement with dance has the potential to help them not only acquire skills but also positive social values. If an all-consuming interest is only in targeting awards or staging superlative performances, then dance as a creative, enjoyable, and meaningful educational experience is inevitably compromised. Young people can gain more from dance if an excessive preoccupation with judging outcomes does not form the bulk of their experiences. School dance CCA programmes should be developed primarily to open multiple possibilities, giving students opportunities to creatively experiment with dance, exploring different styles and nuances. Much can be done to promote the role of dance within the schools and community. In recent years, both the MOE and NAC have worked diligently to elevate the scope of dance in Singapore. Currently there is an emerging prominence and clearer understanding of dance as an art form. Such efforts at promoting dance must continue so that dance education may achieve wider recognition, support, and acceptance in Singapore.

Concluding remarks

Singapore provides many opportunities for young people to engage in dance. However, significant dance educational experiences for the tens of thousands of young people should extend beyond excelling in skills, technique, and performances and competitions. Local dance educators can empower young dancers and facilitate growth in the artistic and creative process, raising awareness that learning in, about, and through dance is a sustained, long-term process.

The schools and general population could become more aware of the value of dance to education and the effects that this will have on the social-cultural and economic capital of Singapore. Dance advocacy, research, and policy review and curriculum development require a stronger focus, along with human resources and material and funding support for youth dance development.

If youth-centred learning is about holistically cultivating kinaesthetic, aesthetic, cognitive, and psychological-social growth, then making dance youth-centred means putting young people at the forefront and using dance instruction to reach them and directly address their needs. Since dance collectives and companies and policy makers exist at the heart of dance education, the responsibility that lies with them is understandably great – they support the youth dance journey, and each of them is capable of bringing about meaningful dance experiences for youth at every level.

As Peter Brinson (1991) says, 'Mastery of the body and its use for expressive purposes develops self-confidence with great psychological benefits. It encourages and releases original achievement which comes from conscious and deliberate choice'.[1] That dance is more than an important physical and emotional activity makes dance education in the schools and community writ large, and for our young people, all the more important. I look forward to a day when Singapore sees a greater emphasis on dance in arts education, and that with educational policy reforms, dance education and its outcomes become even more fully accessible to young people in the Lion City.

Note: This is a revised and updated version of the original chapter in Burridge, Stephanie and Carino, Caren (ed.) (2014) *Evolving Synergies: Celebrating Dance in Singapore*, Routledge, India.

Note

1 Brinson, P. (1991). *Dance as Education: Towards a National Dance Culture*. Routledge, Abingdon, UK. p. 74.

Bibliography

Brinson, P. 1991. *Dance as Education: Towards a National Dance Culture*. Routledge, Abingdon, UK.

Burridge, S. & Wright, S. 2006. 'Creativity: Representational Practices in Artistic Domains' in *Conference Proceedings at the Hong Kong Dance Festival 2006: Imagining the Future: Dance Education in the 21st Century*. Hong Kong: Hong Kong Arts Development Council, Hong Kong Dance Alliance and Hong Kong Academy of Performing Arts, pp. 64–71.

Chan, C. Y. A. 2005. 'Is Hong Kong Ready? An Imperative for Dance Education' in Mohd Nor, Mohd Anis & Murugappan, R. (eds.) *Global and Local Dance in Performance*. University of Malaya, Kuala Lumpur, Malaysia: Cultural Centre, University of Malaya: Ministry of Culture, Arts and Heritage Malaysia, pp. 233–237.

Dyson, J. 2006. 'A New Model for Dance Education in Australia' in *Conference Proceedings at the Hong Kong Dance Festival 2006: Imagining the Future: Dance Education in the 21st Century*. Hong Kong: Hong Kong Arts Development Council, Hong Kong Dance Alliance and Hong Kong Academy of Performing Arts, pp. 95–101.

14

INTERCONNECTIONS

An overview of contemporary dance in Singapore from 1990 to 2020

Melissa Quek

My aim for this chapter is to use descriptions of the work of some key choreographers and their interconnecting relationships to sketch an impression of Singapore's contemporary dance scene from the late 1990s to the present. This is not a history of modern and contemporary dance in Singapore; however, I would like to acknowledge the influence of seminal figures such as Goh Lay Kuan and Lim Fei Shen. So, with such an array of independent choreographers presented in one programme on 12 October 2012 at the Esplanade Theatre Studio as part of the 'Dedicated to You' series of performances that celebrated the Esplanade's 10th anniversary, *Side by Side* was propitious and is the starting point for this chapter. It was billed as presenting 'five established independent dance choreographers who have made an impact on Singapore's contemporary dance scene' – they included Tammy Wong, Joavien Ng, Daniel Kok, Ming Poon, and Scarlet Yu (Side by Side Video archive).

From Wong's *Andante* and her past works, it was clear that she has a knack for choosing rhythmic music that audiences identify with – Bach, Bobby McFerrin, American folk tunes, and hymns. It is music that gets them swaying in their seats or evokes deep memories. This was something different at a time when much of the work in the Singapore dance scene was minimalist or focused on developing and presenting a concept. Trained in the United States, Wong had a modern dance aesthetic; narrative and a strong emotional motivation ran through most of her pieces. She would remark that, just as an author is told to 'write what you know', she choreographed from the subject she understood the best – herself.

Another work in the Esplanade *Side by Side* was *My Superhero* by Joavien Ng. The words, '[m]y journey' stand out brightly on a black background. Ng's voice is heard over the speakers: 'the day I was born I knew there was something special about me'; a body swathed in silver-white lycra from head to toe lies limp on a chair. There is eerie, haunting music and ethereal voices echo. And then a photograph of a baby in a little frilly dress is projected. The amorphous silver-white being goes to a clothes rack and puts on a red bra, navy blue underwear, a bright red tube top, and a thick black belt; struggles into black boots; reveals her long black hair; pulls on a yellow headband and wrist bands; and poses on a pedestal. To the blaring disco music of the Wonder Woman theme song and a wall of flashes behind her,

DOI: 10.4324/9781003160007-24

she spins, hips bouncing as she makes multiple rotations. Just like that, she has transformed into Wonder Woman. Her next iteration comes after she has shared about her childhood pet dog. She has her arms punching straight into the sky and attempts to balance on one leg, giving up she settles for a lunge, her arms splaying out behind her as she lifts the cape in demonstration of her flight; hand fisted to her mouth, it plays the trumpet as she hums the heraldic tune of Superman. This elicits a titter of laughter from the audience.

While the autobiographical appeared in the emotional content of Wong's dances, independent artist Joavien Ng's work is more overtly so. For instance, in *Body Swap*, performed by Ng and co-choreographer Dani Brown, 6–7 June 2009, as part of the Singapore Arts Festival at the Esplanade Theatre Studio, Ng literally lived another woman's life by travelling to Hamburg to inhabit her house, meet her friends, and live with collaborator Dani Brown's boyfriend. In fact, in 2011, while sharing a hotel room in Seoul with Ng (we were there for the Seoul International Dance Festival), over cups of yoghurt, we discussed her belief that she could genuinely be self sufficient: all she would need was a goat and two chickens. She really did need to have her eggs every day. A year later, this philosophy was born out on stage in *A Life Performance*, a solo by Ng, 10–12 May 2012, at the Esplanade Theatre Studio, with giant cut-outs to represent all she would need for her self-sufficient life. This performance, like much of Ng's work, was a personal and poignant combination of the surreal and absurd that used humour to discuss contemporary Singaporean issues such as sustainability and the space crunch. There was minimal dancing, but the movement vocabulary was very specific. A particularly beautiful moment was when Ng held a cut-out of a horse's head in front of her and a shark's tail behind. As she slowly rotated, this fantastical merhorse became a symbol of imagination and an ability to adapt. This same whimsy is apparent in Ng's *Lab* (September 2008), a duet with Ricky Sim at the Esplanade Theatre Studio about living in a controlled habitat where they appeared like lab mice being studied as they responded to their changing environment, whether the introduction of another being or a new toy (Lab footage).

Although they danced together in *Lab*, choreographer Ricky Sim had taken a very different route from Ng. Like Kuik Swee Boon, artistic director and founder of T.H.E, Sim had been a company member of Singapore Dance Theatre for several years before he left to do his masters at Queensland University of Technology – after his return, he presented *Transparent Ground* at the Substation Theatre for the M1 Fringe Festival 2006. Among the local cast members were Kuo Jing Hong (daughter of theatre pioneer Kuo Pao Kun and dance doyenne Goh Lay Kuan), Albert Tiong, and me. Sim was exacting and serious inside the studio and pushed for nuanced details and musical exactitude. This same demanding musicality has continued into his subsequent works. After that performance, Sim continued to give a platform for development to young dancers with a desire to dance. He opened his dance studio and company Moving Arts. The company began to provide a scholarship-training programme for teenagers and young adults, and out of this his performance company, RAW Moves was born in 2011. As an amateur–professional company, it filled a niche in the Singapore dance scene, as it is was made up of full-time dance practitioners and professionals engaged in other fields. RAW Moves became part of the National Arts Council's Major Company scheme in 2019 and streamlined to three to four full-time performing members and chose to focus more on experimental work and collaborations with artists from other disciplines.

In contrast to this, Frontier Danceland is one of the first contemporary dance companies to be established in Singapore. Set up in 1991 as a Chinese modern dance company by artistic director and choreographer Low Mei Yoke, it started to take on new life when

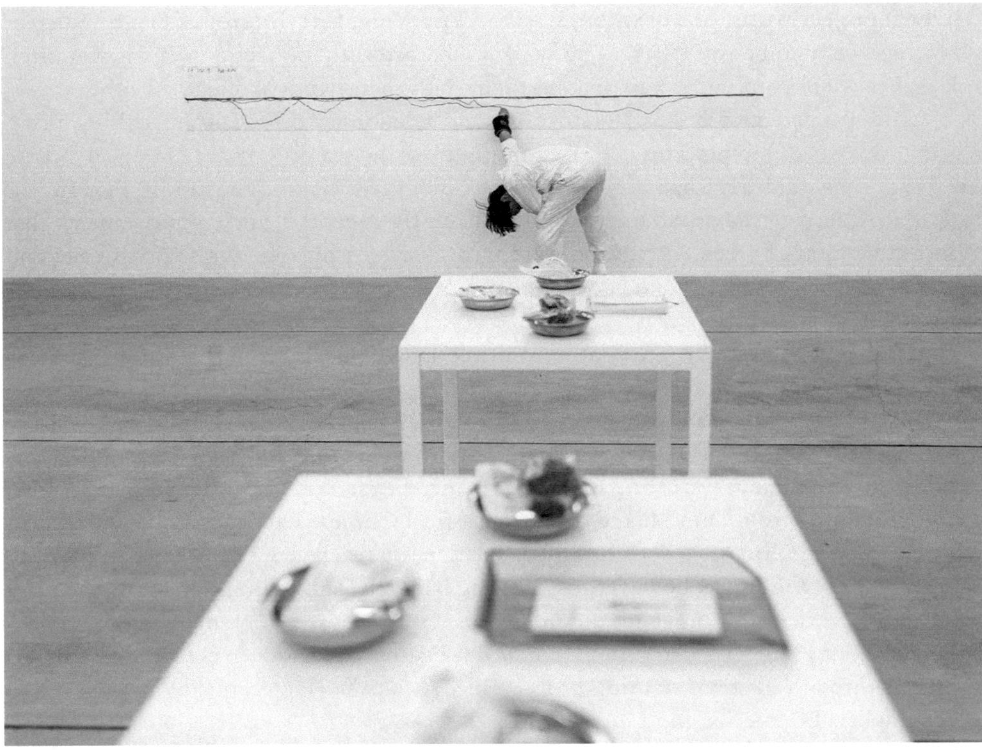

FIGURE 14.1 *Being, and Organs* (2019)

Source: Conceptualized by Paul Gong in collaboration with artistic director Ricky Sim

Dancers: Matthew Goh, Pichmutta Puangtongdee, and Stephanie Yoong (understudy)

Photographer: Crispian Chan (CRISPI) for RAW Moves

it first converted one of its studios at Stamford Arts Centre into a performance space. This was the beginning of their 02 Square series. In the dance drama *Something Wrong* conceived by director Nick Ng (11–14 January 2007 at Stamford Arts Centre), Law Kian Yan, Grace Wee, and I hid behind the curtains, played with boxes, and climbed in and out of the windows. For what seemed to be the first time, the humorous side of aforementioned performer and choreographer Albert Tiong emerged as he lip-synced a Hokkien song into a pair of maracas. Soon after, Malaysian-born Tiong became the resident choreographer of Frontier, and he worked with Low to transform the company. He started regular company classes to train the new dancers, such as a then-recent graduate Chiew Peishan, and the company started conducting several performance seasons a year. This included the Dancer's Locker platform that gave the members a chance to hone their choreographic skills. In 2012, Tiong left Frontier to begin his own company, Re:Dance Theatre (RDT); some of the founding members were Dapheny Chen, Seow Yi Qing, and Rachel Lum, who had danced for Tiong while he was still with Frontier.

With the departure of Tiong, Frontier restrategized to become a repertory company, adopting the model of bringing in a different guest choreographer for each of their seasons. Meanwhile, RDT presented its debut work, *Whisper of the Tempest*, 17–18 April 2012, at Goodman Arts Centre (GAC) Black Box, performed by the founding members and additional project artists, one of whom was Foo Yun Ying, first director of Dance Nucleus and formerly a member of T.H.E. The work is characteristic of Tiong. His dancers often look as if they are painting the air around them with their hands and feet. At other times, his work is peppered with small gesture phrases relentlessly repeating: a shaking head, twisting hips, or pumping arms. Possibly inspired by the training in tai chi and Chinese dance that he would have had as a dancer in Cloud Gate Dance Theatre of Taiwan, his work is athletic, and the vocabulary consists of swiftly executed curving movements that are a combination of speed, energy, and strength. Between his time with Cloud Gate and starting his own company, Tiong had also been a principal dancer for Odyssey Dance Theatre (ODT).

ODT's arts education wing was around in the late 1990s, but the first official performance was *Odyssey to the Sublime* by a cast of seven females (including Joey Chua, who later danced for Ah Hock and Peng Yu, AHPY) on 20 January 2000 at Jubilee Hall (Raffles Hotel). Both local and international guest performers embarked on their individual Buddhist-inspired quests for calm. The performance began with a pensive solo and ended quietly with poses from Dun Huang dance (a classical Chinese dance form that was recreated from ancient Chinese cave paintings) as the dancers reached a state of tranquillity. My main memory of the dance, however, is of tossing hair and flailing limbs grabbing the air, a look that was, at that time, in vogue among works with a Chinese contemporary dance aesthetic. In recent years, ODT Artistic Director Dr. Danny Tan has been able to hire dancers developed through his own training programme, such as current company members and choreographers Lo Pui Sze and Linnea Ong.

Another company that has been training dancers in its rather specific style is Kuik Swee Boon's T.H.E. While the main company is full time, the Second Company, aimed at younger dancers, meets twice a week for classes and has a season about twice a year. *Silence from the Other Side of the Moon* would later become a segment of Kuik's seminal da:ns festival 2007 offering, *Silence*, a work that reinforced his movement style, usually described as sinuous and liquid but powerful, and set his feet firmly on the path towards the genesis of his company, many of the founding members having come from this cast. In the darkness, a man sits perched on one of three walls placed in the corner. He begins to sing. A woman walks slowly forward as hooded figures dot the space. As a voice mumbles through the speakers, a combination of small repeated movements and large spiralling motions slash through the space. The dancers are unfailingly fluid, with moments of tightly leashed energy and unbridled violence; bodies are thrown and slide across the floor (Silence Archival footage). Kuik then strengthened his company's repertoire with the addition of resident Korean choreographer Kim Jae Duk and developed and refined the teaching of his movement style. Based on the interplay of mind, body, and heart (emotions) he terms it 'The Hollow Body'.

While T.H.E was looking for space to rehearse, ODT had already been under the NAC Arts housing scheme for some time. So when ODT was preparing to move from Telok Ayer Performing Arts Centre (TAPAC) to the new Aliwal Arts Centre, Linnea Ong (in collaboration with visual artist Tang Ling Nah) created *Of Lanes and Memories* (4–11 August 2012) within the ODT studios (a white room and a black box) as a farewell to the space and their time there.

Another company that paid homage to its space, the Cairnhill Arts Centre, was Singapore's first full-time contemporary dance company, The Arts Fission Company (TAFC), co-founded in 1994 by choreographer Angela Liong and visual artist Chandrasekaran, with

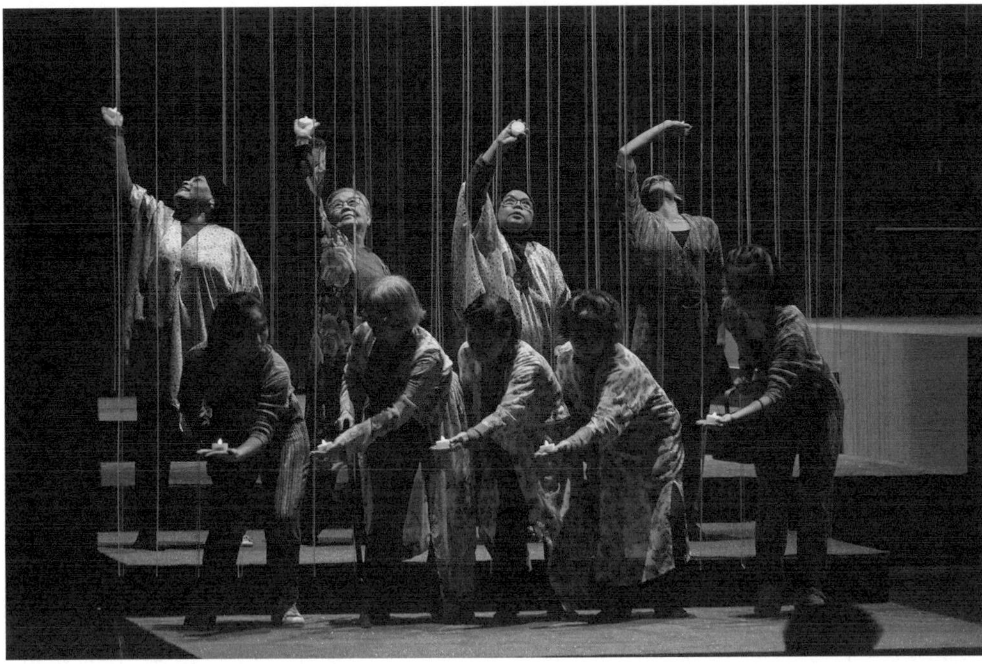

FIGURE 14.2 *garden.uprooted* (2018)

Source: Choreography: Angela Liong

Photo: The Arts Fission Company collection

Folly of a Garden II (26–28 July 2012 at Cairnhill Arts Centre), the second in a trilogy inspired by the auction of 2 of the 12 bronze zodiac animal head statuary from China's summer palace. Described as a bestiary masquerade,[1] audience members were issued masks with their tickets and then brought on a tour of the grounds with a travelling trio of antique Chinese instruments composed by Joyce Beetuan Koh. Against the lush greenery of a hidden cul-de-sac, dancers sported elaborate headpieces, designed by creative designer and co-founder of Ecnad Project Ltd, Lim Chin Huat, that proclaimed the animal they represented. One of the greatest strengths of Arts Fission is its commitment to site-specific work. In fact, multi-disciplinary artist Elysa Wendi, former assistant artistic director of Arts Fission and the first company member when it became full time in 1999, agrees that the signature of the company is 'site-specific works with unusual setting, critical works that comment on current affairs, climate changes and the Asian sensibility towards the world at large'.[2] Even in a more traditional performance space, such as the Esplanade Concert Hall, Liong works in a site-specific manner. The community-based work *The Rite of Spring: A People's Stravinsky* by TAFC, held on 22 June 2013 at the Esplanade Concert Hall, in collaboration with the Philharmonic Orchestra headed by maestro Lim Yau, celebrated the 100th anniversary of Stravinsky's scandal-inducing composition. Apart from the main company members, the cast consisted of young dancers from Arts Fission's own education programme, volunteer dancers including students from LASALLE College of the Arts, and veterans from two senior centres. The dancers skilfully manoeuvred in the small space in front of the orchestra and along a small raised platform behind. In the area where the choir would usually be seated were the senior performers with their prompters. In their elaborate costumes and head-dresses, they became a beautiful performing installation that framed the rest of the action. This work was followed two years later by *The Mazu Chronicle* (2015), and in 2018, *garden.uprooted* completed this trilogy of collaborations between Lim and Liong for seniors, company members, and young dancers from LASALLE.

Another collaborator on these works was Lim Chin Huat, who was working on the costumes and prop design on *The Rite of Spring* the year he closed down ECNAD, formerly Dance Dimension Project, a company he co-founded in 1996 and renamed ECNAD in 2001 with Tan How Choon. *Reborn/FullStop* (4–6 April 2013) was the company's farewell performance held at the Arts House Play Den. The performance showcased costumes and excerpts from past works, one of which included a woman dressed in a long gown; the hem, suspended from the ceiling, would billow out each time she advanced and quickly retreated. This company gave amateur dancers a platform to train and perform, so it was fortuitous that when it closed, Sim's RAW Moves stepped up to the plate. Subsequently, in 2016, the Malay dance–based P71sma, founded by Hasymah Harith, Md Hariz Bakri, and Norhaizad Adam, which builds contemporary ideas and concepts upon its traditional Malay dance training, provided a welcoming environment for professional, pre-professional, and amateur performers to find expression for their art.

The collaborative way of working that she experienced at TAFC fed into Wendi's practice as well. This is not surprising, since she was with the company from 1999–2009, and she notes Liong's connections to music, literature, and food as an inspiration. Her contemporaries at TAFC were Scarlet Yu, Aaron Khek (Ah Hock), Ix Wong, Bobbi Chen, Roslina Bte Yusoff, and Wong Wai Yee (a founding member of the company). After leaving Arts Fission, Khek and Wong started AHPY, a contemporary art collective that was very active in Singapore until it moved to straddle both Singapore and Malaysia. Yu and Wendi subsequently relocated to Hong Kong. Shortly after striking out on her own, Wendi had started her first collaboration with a film-maker in 2009, although her first film was created in 1999 with a visual artist. Since then, Wendi has been intrigued by the hybrid possibility of dance and film (Elysa Wendi, email interview).

She works on 'presenting dance in relation to different settings, mediums and methodologies in contemplating the "process" of choreographies as a per-formative substance' (Elysa Wendi, email interview). Therefore, although Wendi is interested in a variety of mediums, screendance in particular lends itself to these explorations because, according to her, although screen recording may have killed the ephemeral beauty of depicting presence, in the absence of live-ness, the manipulation of technology brings a new concept to presence. This is why Wendi has become known in the Singapore scene for her screendance works. One of the first of these to tour the film festival circuit was *Void Decked* (2010), created with director Azhar Syukor and performed by former Arts Fission dancer Roslina Bte Yusoff at a void deck near Wendi's house in Tiong Bahru. Motivated by the lack of resources, opportunity, and collaborative conversations about the medium, she co-founded Cinemovement in 2015 with film producer Jeremy Chua as a platform for filmmakers and choreographers to 'explore the notion of contemporary dance/film through cultural exchange in a laboratory setting' (http://cinemovement.sg).

Cinemovement in particular has been instrumental in furthering exploration into screendance among Singaporean practitioners. Norhaizad Adam of P71sma attended the first two Cinemovement laboratories in 2015 and 2016, and subsequently forays into this

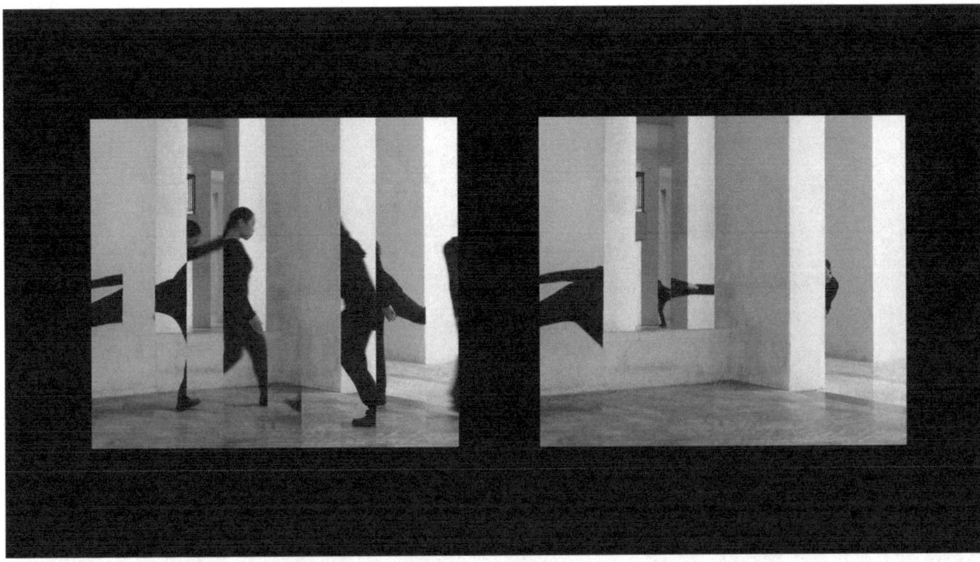

FIGURE 14.3 *Void Decked (the amended credits)*

Source: Created by Azhar Syukor and Elysa Wendi

Performer: Rosalina Bte Yusoff

Photographer: Azhar Syukor

form can be found among the company's works. Independent artist Chan Sze Wei, who attended the Cinemovement Lab in Macau in 2018, where she made short films such as *Chickadee*, credits it as a significant turning point for her (Chan Sze Wei, email interview). Interacting with professional film directors and producers gave her new skills in planning and collaborative concept development that enabled her to transition from abstract movement-based shorts to making a dance documentary with Looi Wan Ping (Chan Sze Wei, email interview). Looi is a cinematographer and editor whom Chan met at Cinemovement. With her new perspectives on spectatorship and the screen as a medium, she started making interactive video installations, and projections and live feeds are often included in her performance works.[3]

Similar to Liong's affiliation to creating work in series, Singapore-born, Berlin-based choreographer and performer Ming Poon has been presenting his work in developmental phases, building on the audience reactions to one work to create the next. In his two-year residency, begun in 2011, as an associate artist of the Substation, he began researching the lives and experiences of people living with HIV (PLHIV) and the silence with which Singaporean society surrounds this topic. His work in this area goes beyond revealing the pain of those he has interviewed. In fact, in dialogue with him, it is clear that he resists labels and categorization. He is not comfortable with terms such as 'subject' and 'researcher'.[4] Given that the work is as much about PLHIV as it is about the avoidance tactics and strategies of silence employed by Singaporeans, it would be hard to decide who the real subjects of his research are. Poon does not pretend to be a champion of a cause but instead presents his body and performance as a product of his interactions with research and the research process. In *(un)it: HD85828 | in.ViSiBLE* by Ming Poon (20 December 2013), held at the Substation

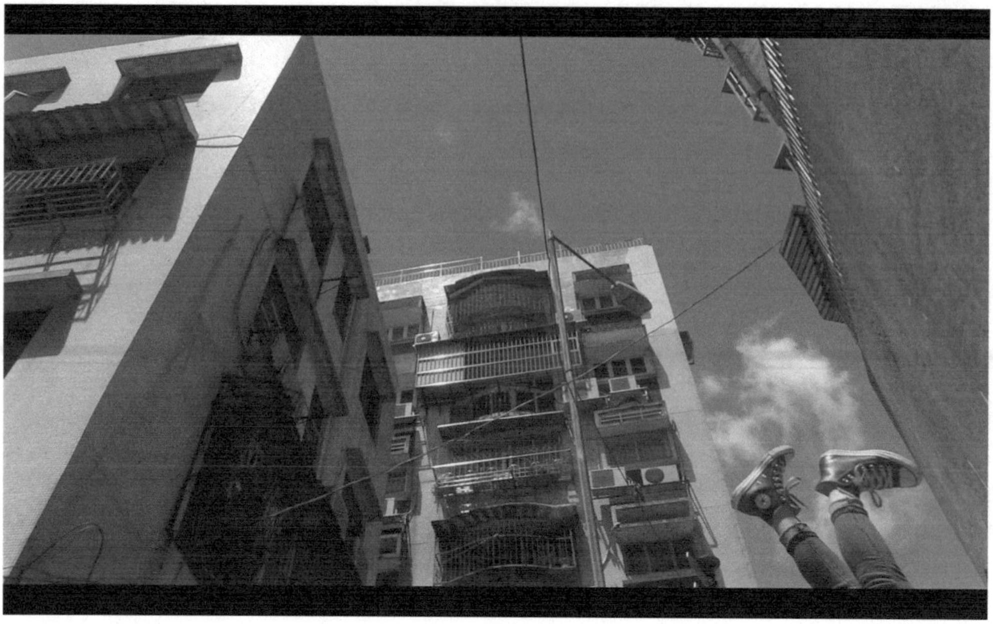

FIGURE 14.4 Chan Sze-Wei

Source: Film shot *Chickadee* (2018)

Photographer: Gladys Ng

Theatre, one of the main conundrums he faced was how to perform the invisible and to preserve privacy without hiding. In fact, when inviting me to watch the show, he said 'there is nothing to see'. In the Substation Theatre, with the seating area pushed back, a hundred or more numbered A4-sized papers lined the walls of the room, a microphone in each corner, and light bulbs dangled from the ceiling to lie on the floor. Poon stood covered in a white hazmat suit revealing only hands and face. In the stilted and angular manner of a robot, he directed the audience to sit around the performance space, delivering his instructions clearly. He asked for volunteers to stand in place and hold up the light bulbs, and he did not move on until every bulb was held. The remaining audience members, some of whom had contributed the text, were told to pick a numbered paper, and when it came to their turn to read the text printed on the other side. They were to keep picking numbers and reading until the bell rang, signalling the end of the hour-long performance. This text was a selection of excerpts from the interviews Poon had conducted and collected; it became a symphony of voices intoning the words of pain, hope, rejection, love, actions, and inaction. If a microphone was in use, some of the light bulbs would switch on, but if there was silence, the performance would continue in darkness. The more people spoke, the brighter it became, revealing more of the performance. Poon moved with his usual lithe grace, sometimes drawing close to the light, sometimes hiding. At some point, he had discarded his clothes and was completely vulnerable in his nakedness, nearing the volunteers or smoothly skirting round them; in one heart-rending moment, the tension and frenzy built as he desperately banged himself against the floor. And, all too soon, the performance ended.

If Ng's work for *Side by Side* had the audience in thoughtful chuckles, Daniel Kok's work *The Cheerleader* (2012) had them guffawing. In walks, Low Kee Hong, former general manager of the Singapore Biennale and Singapore Arts Festival, dressed in a white suit with red sporty stripes down the side. The suit is grotesquely distorted, as it is tightly packed with intestine-like tubes that bunch around his stomach and hips but escape through a slit in the front to trail along the floor. Holding two black pom-poms that he promptly puts down, Low sharply executes a series of cheerleader-like moves, arm gestures, star jumps, leaps, and twirls across the stage to end in a lunge, arms lifted in the air spread in a triumphant 'v' shape. He holds for applause. Picking up his pom poms, he starts a chant, 'Go Kee Hong, say Low Kee Hong, High Kee Hong, love Kee Hong', then pulls out the intestines further. After a short interlude about the work being too 'chim' (difficult to understand), Low starts running and trying to high-five the audience members to get the crowd going. He becomes Singapore's cheerleader, but is chanting 'Knnbccb' (an abbreviation of an impolite Hokkien phrase) and presses the crowd to join him. As Kok has written, Low the Cheerleader is relying on 'little more than his body and voice to invite the community to coalesce'.[5] Kok's comfort with popular dance forms seems to come from the 'hodgepodge of classes' in a variety of forms that he took in different times and places in his life. Although he was initially trained as a fine artist, it seems that Kok has always had an affinity for dance and performing. Kok, who is the second artistic director of Dance Nucleus, like Liong, enjoys working with experts from different (non-arts) fields. Where Liong mines data from these different fields, Kok invites an economist, a sociologist, or a psychologist into his working process as a consultant. Both have therefore taken interdisciplinary collaboration beyond the closed circle of artists to make much larger leaps. Kok's research continues to centre on questions of relationality, spectatorship, and audienceship (Daniel Kok, email interview).

In 2017, for the Singapore International Festival of the Arts (SIFA), in the final moments of *Mark*, the performers huddle together on a giant sheet of paper throwing coloured powder into the air. The vibrant puffs disappear as they are wafted away, carried by the wind, a testament to the temporality of performance. Remaining are the traces of charcoal, chalk, and powder collaboratively drawn by audience and performers, who are now panting from their exertions and covered in pigment. I stand among them, marked by this experience of dancing the set score of collective drawing, where we had to be constantly aware of our shifting perceptions between imagination and the physical realities. *Mark* was performed in three different public spaces over four performances by nine dancers. Of these performers, several, such as Jereh Leong and Pat Toh, subsequently became associate artists with Dance Nucleus, continuing to work with Kok to refine and articulate their individual practices.

Dance Nucleus was an initiative begun in 2015 by NAC to provide a dedicated space for the development of the practices of independent dancers and choreographers. Initially run by Foo Yun Ying, Daniel Kok took over as artistic director in 2017, and with its official launch in January 2018, it was clear that Kok's own experiences and approach to being an independent artist had greatly informed the design of the programmes. There was a greater focus on practice-based research and overall knowledge production for independent contemporary performance and performance-makers. The discussions and talks acknowledge the wide range of skills and concerns of independent artists that are different from performers/choreographers working within a company structure. Speaking of her experiences as an associate artist with Dance Nucleus, Toh said that 'the development had been quite

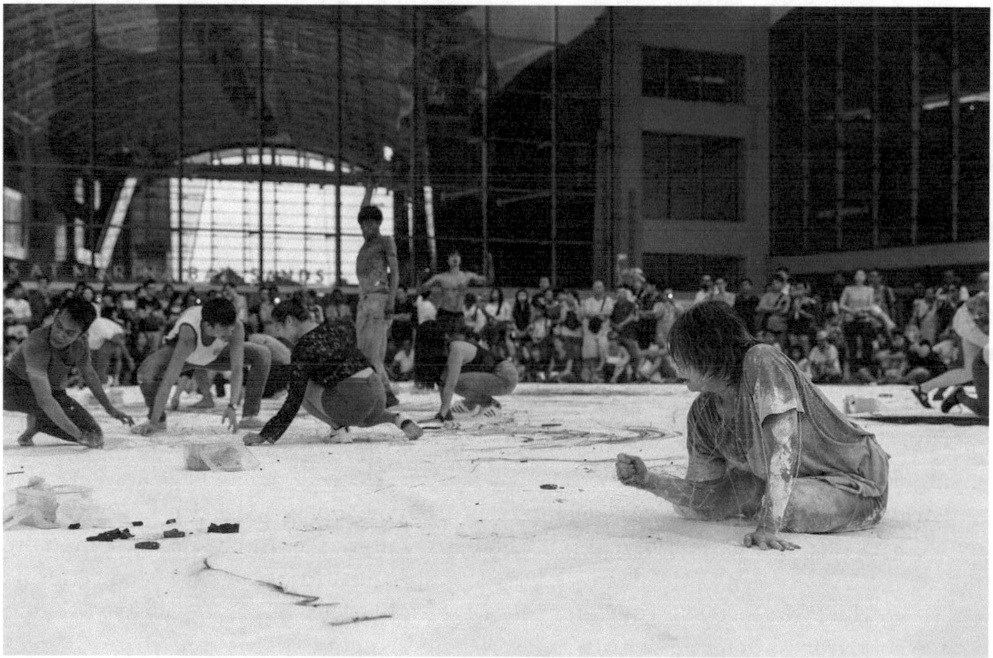

FIGURE 14.5 *Mark* (2017)

Source: Commissioned by and premiered at the Singapore International Festival of Arts (SIFA) 2017

Choreographer: Daniel Kok

Dancers: Lee Mun Wai, Melissa Quek, Patricia Toh, Jereh Leong, Yazid Jalil, Felicia Lim, Elizabeth Chan (Singapore), Phitthaya Phaefuang 'Sun' (Bangkok), Otniel Tasman (Surakarta)

Photographer: Bernie Ng

holistic from critical practice, to community building, to the business side of things' (Pat Toh, WhatsApp interview).[6]

Dance Nucleus helped Toh to focus more on the phases of creation as a vital part of her practice. This means that she now takes into consideration the resources required for the 'full depth of creation' starting from conceptualization to research and then performance. She found that Dance Nucleus 'offered exposure to performance theories and practices, it helped me build a strong support network with other independent artists, a critical community (local, regional and international) to think through and develop creative processes with' (Pat Toh, WhatsApp interview).[7] As part of her residency with Dance Nucleus, Toh had the opportunity to work closely with Ming Poon to refine her thinking and approach to her sports and fitness-inspired movement practice.

As dramaturg Lim How Ngean writes of the independent artists in *Side by Side*, I say of these artists here that they 'represent the commonalities – as well as the diversities – of dance in this country'.[8] Yet this selection is limited to the work of artists with European or American influences and some who have developed from the initial Chinese dance impetus that propelled modern dance in Singapore. New approaches to other traditional dance forms have been on the rise in the Singaporean contemporary dance scene, such as Bharatanatyam in the works of Maya Dance Company and Soul Signature, Odissi as in the contemporary works of CHOWK led by Raka Maitra, and *silat* as in the works of Era Dance Company

As the influences multiply, contemporary dance in Singapore will continue to diversify, but I believe the interconnections of the people in the industry are what shape and propel the dance forward.

Note: This is a revised and updated version of the original chapter in Burridge, Stephanie and Carino, Caren (ed.) (2014) *Evolving Synergies: Celebrating Dance in Singapore*, Routledge, India. The second part has been thoroughly rewritten to reflect the many new developments in the past decade.

Notes

1 www.artsfission.org/#!the-folly-of-a-garden-2/zoom/mainPage/image1nm1 Accessed October 2013.
2 Personal email interview conducted in September 2013.
3 Personal interview, 26 May 2020.
4 A view he expressed in the post-show Question & Answer session, the Substation performance, 5 April 2013.
5 Personal interview, 11 September 2013.
6 Personal interview, 29 May 2020.
7 Personal interview, 29 May 2020.
8 Lim, How Ngean, 'What's in a Tribute?', Side by Side Programme Book, Esplanade Theatres, 2012.

Reference

Lim, How Ngean. 2012. 'What's in a Tribute?', *Side by Side Programme Book*. Esplanade Theatres.

Video archives

Lab. 2008. Joavien Ng (choreographer). Esplanade Theatres. Video archive (accessed in September 2013).
Side by Side. 2012. Tammy Wong, Joavien Ng, Ming Poon and Scarlet Yu, Daniel Kok (choreographers). Esplanade Theatres. Video archive (accessed in September 2013).
Silence. 2007. Kuik Swee Boon (choreographer). Esplanade Theatres. Video archive (accessed in September 2013).

Website

http://cinemovement.sg (accessed in June 2020).
www.artsfission.org/#!the-folly-of-a-garden-2/zoom/mainPage/image1nm1 (accessed in October 2013).

Interviews

Chan Sze Wei, email interview. 26 May, 2020.
Daniel Kok, email interview. 11 September, 2013.
Elysa Wendi, email interview. 13 September, 2013.
Pat Toh, WhatsApp interview. 29 May, 2020.

Taiwan

15

IDENTITY, HYBRIDITY, DIVERSITY

A brief history of dance in Taiwan

Chen Ya-Ping

An island in search of its identity

In a series of debates on the issues of colonial history, national identity, and subjectivity construction in Taiwan published in the literary journal *Chung Wai Literary* (*Zhongwai Wenxue*) from 1995 to 1996, literary scholar Liao Chao-yang proposed a theory of 'empty subject' (*kongbai zhuti*) as the essential nature of Taiwanese subjectivity:

> Emptiness does not mean void; the subject being empty does not mean 'the death of the subject.' . . . The empty subject has to continually 'take in' objects in order to adjust the relations between the inside and the outside, and to maintain the efficacy of the emptiness in the evolvement of concrete historical experience. . . . The empty subject can also be seen as a unit or form of existence . . . a *flow* of life experience and a *space* rather than content. In other words, emptiness does not exclude content; rather, it is a *space* which takes in, alters, and gives meanings to content.
>
> *(Liao 1995: 119, italics added)*

The concept of 'empty subject' as defined here not only points out succinctly the fluid nature and continually formulating and reformulating state of identity in Taiwan caused by the complex history and geopolitical location of the island but also captures vividly the anxiety over identity problems which has troubled the Taiwanese for more than a century.

Located approximately 100 miles off the southeast shore of the Chinese mainland, Taiwan occupies a strategic position on the oceanic route between East Asia and European countries. The original inhabitants of Taiwan are diverse tribes of Austronesian peoples, now referred to as the Taiwanese indigenous people (*yuanzhumin*). Though Chinese expeditions reached Taiwan as early as the 3rd century, the island was not extensively settled by the Han Chinese before the second half of the 17th century. In 1662, Koxinga (Cheng Cheng-kung) and his followers, retreating to Taiwan to continue their mission of restoring the Ming Dynasty (1368–1644), laid down the institutional, social, and cultural foundations of Chinese civilization on part of this frontier island. Although the Qing court (1644–1912) formally annexed

DOI: 10.4324/9781003160007-26

Taiwan in 1683, the island remained a troubled frontier characterized by periodic revolts and frequent ethnic or sub-ethnic conflicts well into the 19th century. There was little sense of a collective identity among the settlers but loyalty to regional communities based on same surname, ancestral origin, or dialect. The decisive turning point in the formation of the Han Chinese islanders' identity – from the aforementioned fragmented identities to the collective identity as Taiwanese or Chinese – came in 1895 when Taiwan was ceded to Japan after the Qing was defeated in the first Sino-Japanese War. After several months of bloody fighting with local militias, the Japanese finally took control of the island, and Taiwan entered 50 years of Japanese colonization.

Over the course of the five decades, the Taiwanese were made subordinates to the Japanese in social, political, and cultural spheres. Yet, at the same time, the Japanese also facilitated major investment in education, infrastructure, and technological upgrading and served as the channel for filtering in 'progressive' ideas not only in science and technology but also in arts and culture. The birth of concert dance in Taiwan was a direct outgrowth of this unique process of 'colonial modernization.' Lin Ming-te, Ts'ai Juei-Yueh, Li Ts'ai-e, and Li Shu-fen were among the first Taiwanese going to Japan to study dance with Ishii Baku, his disciples, or other Japanese dance pioneers. The Japanized Western creative dance forms – including Duncan-style free dance, neo-romantic ballet, certain aspects of German *Neue Tanz*, and Oriental dance inspired by Denishawn – constituted the foundation of early Taiwanese concert dance tradition.

Another strand of Japanese colonial legacy in dance came from the northeastern region of mainland China, also known as Manchuria. Bordering Russia to the north, Manchuria had historically a strong Russian cultural and political presence and was colonized by Japan from 1931 to 1945. After the Chinese Nationalist Government (Kuomintang or KMT) was defeated by the Chinese Communists and retreated to Taiwan in 1949, many mainlanders fled Red China with the KMT to Taiwan. Among them were dancer/educators Kao Yen, Li Tien-min, and Liu Feng-hsueh, who played seminal roles in laying down foundation for higher education in dance in Taiwan. Li and Liu were both born and educated in Manchuria.

As part of the agreement in the Cairo Declaration (1943), Taiwan was retroceded to the Republic of China (the R.O.C.) ruled by the KMT after Japan lost the Second World War. Once again, the island became war booty, and its inhabitants were forced to undergo another identity change – from Japanese back to Chinese. As part of the KMT's scheme of re-Sinicizing the islanders, efforts in promoting '*minzu wudao*' (Chinese ethnic or national dance) were inaugurated around 1952, spearheaded by the General Political Department under the Ministry of Defence. In the *Minzu Wudao* Movement, hybridity of forms and concepts was part of the creative strategy practiced by many of the dance pioneers to fulfil the official agenda of anti-communism and Chinese nationalism. Ts'ai Juei-Yueh and her Taiwanese peers blended the techniques and imaginative creativity they had learned from the Japanized Oriental dance with Chinese ethnic gestures they had gathered to choreograph the many Chinese ethnic dances they had never seen (Chen 2008).

Later, another kind of hybridity was carried out by Liu Feng-hsueh and Lin Hwai-min for the purpose of formulating a modern Chinese form of dance. Beginning in the late 1950s, Liu experimented on her students with a hybrid style that incorporated Chinese movement motives taken from Peking Opera and other sources with Laban-based modern dance. Liu founded the Neo-Classic Dance Company in 1976 (Lu 2011). In a similar spirit, Lin Hwai-min, the founder and former artistic director of the Cloud Gate Dance Theatre (1973–1988, 1991–), blended Graham technique and Peking Opera vocabulary to choreograph dances inspired by traditional Chinese theatre, legends, or literature. In the late 1970s, when 'Taiwan consciousness', as opposed to the official ideology of Sinocentrism, emerged in the form of the Nativist Cultural Movement (Xiangtu Wenhua Yundung) in Taiwan, Lin integrated Graham technique, martial art movements from Peking Opera, the American avant-garde's practice of outdoor training, and the concept of ritual theatre, as well as the Nativist aesthetics of glorifying labour and indigenous colour, to create the now legendary *Legacy* (1978), an epic dance drama depicting Han Chinese settlers' immigration to Taiwan in the 17th century (Chang and Lin 2005).

Shifting identity and diversification of dance voices

The dance scene in Taiwan entered an era of diversification in the late 1980s and 1990s, coinciding with democratization in politics and increasing internationalization in the culture of the society at large. The appearance of a dance company like Taipei Dance Forum (TDF), founded by Ping Heng one year after Cloud Gate's temporary folding in 1988, was significant. As indicated by its title, the company at its initial stage operated as an open forum, which provided space and professional dancers for young choreographers to present their work. Many important post-Cloud Gate-generation choreographers, who had just returned from graduate study in the United States, started their career with TDF. During this period, several phenomena were eminent in shaping the landscape of contemporary dance on the island: the impact of German *Tanztheater* and the awakening of feminist consciousness, the emergence of 'Eastern body aesthetic dance' (*dongfang shenti guan wudao*), the influence of certain strands of American post-modern dance, and the going independent of first-generation Cloud Gate dancers, among others.

From the second half of the 1980s to the early 1990s, the continuing formation of a Taiwanese subjectivity was accompanied by a process of political awakening. In 1987, the four-decade martial law was finally lifted. The 'rediscovery' of Taiwan's traumatic post-War history, such as the 'February 28th Incident' in 1947 and the 'White Terror' era (1949–1991) of political persecutions, endowed the Taiwanese consciousness with a sense of anti-political authority and/or anti-Chinese colonialism mentality. In addition to this heightened historical awareness of Taiwan's recent past, the rise of feminist consciousness in the society also constituted a significant part of the broader political awakening phenomenon. Inspired by Pina Bausch's *Tanztheater* and stimulated by the women's movement at the time, Tao Fu-lann exposed the constructed and hence performative nature of gender roles in her dance. In a similar spirit, Hsiao Wo-ting examined and protested against the victimized status of women in the country's political history as well as the commodification of women's bodies in Taiwan society.

If the politically conscious dance theatre works represented a desire to construct connections with Taiwan's history and social reality, the emergence of Eastern body aesthetic dances in the late 1980s signified Taiwanese choreographers' ambition to redefine the relation between Taiwan and the Western world in choreographic terms. Rather than adopting Western dance forms, as had been conventionally practiced, the Eastern body aesthetic choreographers set up the goal of conquering the world stage with dance languages inspired by Eastern body-mind cultivation systems such as yoga, *tai-chi*, and the concept of *chi*, among others. Choreographers working in this vein were Lin Hsiu-wei, Liu Show-lu (1949–2014), and Tao Fu-lann.

On the other hand, certain strands of American post-modern dance, most notably contact improvisation, were introduced to Taiwan in the early 1990s. Ku Ming-shen has been actively involved in organizing workshops of contact improvisation as well as choreographing works with various forms and degrees of improvisational technique since then. Originally inspired by Eastern philosophy and concept of movement, mainly the principle of yielding and not forcing, contact improvisation's influence on Taiwanese dance testifies to the reciprocal interaction between the East and the West in contemporary dance. It enriches the expressive dimension of Taiwanese dancers' bodies by enhancing sensibility to one's self and others, in addition to adding flexibility to both the mind and body.

Between 1988 and 1991, Cloud Gate was on hiatus due to financial and other reasons; as a result, many of its first-generation dancers set out to pursue their own dance dreams by establishing new dance companies. In addition to Lin Hsiu-wei's Tai-gu Tales Dance Theatre and Liu Shaw-lu's Taipei Dance Circle, four other Cloud Gate veterans, Lo Man-fei (1955–2006), Cheng Shu-gi, Wu Su-chun, and Yeh Tai-chu, founded the Taipei Crossover Dance Company in 1994, a forum for collaborations with artists from other fields.

Traditions re-invented

For more than half a century, Liu Feng-hsueh, alongside her passion for modern dance creation, has scrutinized ancient Chinese scripts, drawings, sculptures, and notations for the task of reconstructing ancient Chinese dances. From the 1960s to the 1990s, she conducted research in Japan, Korea, and China to further her investigations.[1] Since the early 1990s, she has successfully staged numerous reconstructions of court dances from the Tang Dynasty, demonstrating to modern audiences the solemn and refined beauty of Chinese classical dance. Another attempt to re-create ancient dance tradition is the formulation of the *Nanguan* dance form, which concurred with the resurgence of interest in *Nanguan* (southern wind) music in the past three decades. Cloud Gate veteran Wu Su-chun was instrumental in the formulation of the new *Nanguan* dance theatre, which utilizes modern theatrical means and senses to highlight the exquisite movement vocabulary derived from the Quanzhou Opera and string puppet theatre in the southern Fujian province in China, resulting in the 'new' classical theatre *Nanguan Yuehwu* (*Nangan* Music and Dance). Representative groups of the genre include Han Tang Yuefu and Gang-a-tsui Theatre.

With the resumption of cross-straits exchanges between Taiwan and mainland China in 1987, *minzu wudao* enjoyed a renewed interest among some Taiwanese dance practitioners due to contact with and importation of Chinese 'classical' and 'ethnic' dance styles on the mainland, a 'reinvented tradition' of post-1949 China. The Taipei Folk Dance Theatre, founded by Tsai Li-hua in 1988, produced dances drawing upon this *new* Chinese dance tradition as well as Taiwanese folk dances Tsai gathered and learned through her fieldworks, most noticeably staged religious processions adopted from temple festivities in Taiwan. A similar re-orientation toward Taiwanese folk elements and local culture is observed in the recent development of another *minzu wudao* company, Fevervine Dance Theatre. It has drawn upon the local stories and historical figures of the ancient city Tainan, where the company is located, in site-specific creations in recent years, turning itself into a signature performing art group of southern Taiwan.

Veteran innovators and new talents

After its hiatus from 1988 to 1991, Cloud Gate entered a new stage of development marked by an increased identification with Asian culture and simultaneously a globalized marketing strategy that actively promotes the company on international dance stages. Since the mid-1990s, Cloud Gate has formulated a unique training system for its dancers, which includes Qi Gong (an ancient breathing and energy cultivating exercise), meditation, internal martial arts, modern dance techniques, and ballet. The result is a series of masterpieces, such as *Moon Water* (1998), *The Cursive Trilogy* (2001, 2003, and 2005), *The White Trilogy* (1999, 2006), *Water Stains on the Wall* (2010), and others, which effectively capture the spirit of modernist abstraction while echoing the philosophical and aesthetic essence of traditional Chinese high arts (Chen 2018; Lin 2018).

For 15 years from 1995 to 2009, Lin Lee-chen patiently awaited the fermentation and fruition of her 'Trilogy of Human, Earth and Heaven', *Mirrors of Life* (1995), *Anthem to the Faded Flowers* (2000), and *Song of Pensive Beholding* (2009), works that embrace extraordinary sense of inner corporeal awareness. Walking, ranging from sustained and subdued rhythm to combustible and aggressive attack, serves as the central leitmotif of stage action in all three dances. The total commitment of dancers' minds and bodies to this seemingly simple act, as well as their absolute devotion to carrying out all details of movement, transforms theatrical presentation into ritualistic enactment and points towards an experience of time and space alternative to modern theatre under Western aesthetic influence (Chen 2020).

Another example of intensive dialogue between the East and the West is the work of Yang Ming-lung. A former member of the Trisha Brown Dance Company, Yang has a deep understanding of the philosophical and anatomical dimensions of release technique, which he uses to deconstruct and reconstruct the codified movement vocabulary of Taiwanese hand puppet theatre and Peking Opera. The results are the 'Eastern Current Trilogy' choreographed between 2002 and 2005 and the follow-up *Eastern Tale*, which premiered in 2011. Drawing upon the physical stylization and dramatic energy of traditional theatres, Yang forsakes conventional narrative structure and role playing in favour of exploring such abstract concepts as tension, speed, and structural composition.

In contrast to the creative returns to traditional or primal sources of Taiwan's cultural heritage just described, some dance companies and choreographers are more inclined towards urban sensibility. The Taipei Crossover Dance Company (1994–2010) often explored the life experience and emotional world of middle-age urban dwellers, befitting the mature stage of its four founders' dance careers. Other companies working in a similar mood of metropolitanism include the Century Contemporary Dance Company, founded by Yao Shu-fen, and the Sun-shier Dance Theatre, co-established by Wu Pi-jung and Chang Hsiu-ping. Yao is known for her witty choreography, which draws upon popular cultural elements, and dramatic acting, as well as innovative use of props and sometimes urban landscapes. As for Wu and Chang, the gender role and social/self images of urban women are their primary subject of satirical interrogation.

The founding of the Cloud Gate 2 Company (1999–2019) by Lin Hwai-min and Lo Man-fei was a conscious step in nourishing upcoming talents in dance. For two decades, the company provided opportunities for some of the most promising young choreographers in Taiwan, including Bulareyaung Pagarlava, Wu Kuo-chu (1969–2006),[2] Cheng Tsung-lung, and Huang Yi, among others. Serving the artistic directorship of the Cloud

Gate 2 from 2014 to 2019, Cheng is a versatile choreographer demonstrating mastery over diverse styles of choreography, ranging from minimalist structuralism, to colloquial pastiche, to robust primitivism. In recent years, he has explored and successfully transformed the corporeal culture and spiritual potential of Taiwanese temple fairs and folk rituals into contemporary dance vocabulary and theatrical presentation. In 2020, Cheng succeeded Lin Hwai-min as the new artistic director of the Cloud Gate Dance Theatre, officially bringing this first modern dance company in Taiwan into a new era (Chen 2021).

Another choreographer inspired by the corporeal memory and sensibility of Taiwanese colloquial and folk culture is He Xiaomei, who founded MeimageDance in 2010. Her latest version of the *New Paradise of Silent Island* series (2018) turned the space of a black-box theatre into the energy whirlpool of a temple parade, in which the audience became immersed participants and witnesses of an exciting ceremony and its hidden conflicts. With keen observation of gender roles in contemporary society, He is also recognized for imaginative dance images probing the possibilities of theatrical representation and its interaction with social role-playing.

In 2005, an all-male dance company M Dans (later changed to 'HORSE') was founded by five young men aged between 22 and 27, Chen Wu-Kang, Chou Shu-yi, Huang Yi, Su Wei-chia, and Yang Yu-ming. Working collaboratively as dancers and choreographers, they represent a new phase in the development of Taiwan's contemporary dance. Having absorbed highly diverse dance techniques and styles thanks to the increasingly international approach in Taiwan's dance academies as well as the convenient access to global dance resources over the internet, this new generation move freely and adeptly between different disciplines of arts – dance, theatre, photography, film, and installations. In recent years, Chen and Su remain as core members of the company, while Chou returns periodically to collaborate.

Chen Wu-Kang initiated a research and performance project with Thai choreographer Pichet Klunchun in 2016, in which the two look back at their own corporeal histories as dancers/choreographers and engage in verbal and physical dialogues on the social, cultural, and aesthetic dimensions of their respective genealogies. In the same spirit of choreographic research, Su Wei-chia has devoted himself to the ten-year project of the *Free Steps* series since 2013. In each of the dance episodes, which take place in difference formats and sites, he imposes strict restrictions such as spatial limit, dynamic quality, and body parts to explore the infinite possibilities and nuances embedded in the minute physical expression of three-dimensional dancing bodies.

Dance and the new media

Experimentation with performance and technology has been a major area of exploration by dance companies from Cloud Gate to smaller groups and independent choreographers. In Cloud Gate's *Wind Shadow* (2006), a collaboration with New York-based Chinese artist Cai Guo-qiang, multi-media projections and laser beam effects are used to create apocalyptic stage images which evoke memories of human and natural catastrophes in recent world history. In *Water Stains on the Wall* (2010), ever-changing ink clouds projected on the slanted stage floor not only conjure up the abstract poetics of Chinese water ink painting but also inspire philosophical mediation on time and change in the minds of the spectators.

Chou Shu-yi, a precocious choreographer and expressive performer, has worked with many new media artists and has been well known for creating poetic narratives of dance images across different media. In 2019, he collaborated with director Singing Chen to make a highly successful VR dance film, *Go Astray – Afterimage for Tomorrow*, which transports the spectator into an imaginary interstitial time-space between life and death, past and future, with enchanting images evoking sensuous kinesthetic memories. Another young talent is Huang Yi. In 2012, he premiered his duet with an industrial robot, *HUANG YI & KUKA*, a poignantly beautiful partnership exploring choreographic metaphors about need and rejection, intimacy and vulnerability through the paradoxical combination of human precision and poetic robotics. Three years later, he expanded it into an evening-length piece with the addition of two other dancers and projected video images, complicating the narrative about humanity and technology with connotations of surveillance and manipulation (Lin 2020).

A strong female voice in art and technology, Su Wen-chi is one of the first choreographers to work with new media in Taiwan. In a series of her own solo performance, she investigates the future of corporeality and modes of representation in theatre in the era of digital technology. Her radical experimentations with different material mediums and mechanical or environmental installations have expanded the frontier of dance and new media art in Taiwan. Hsieh Chieh-hua, artistic director of the Anarchy Dance Theatre, is another thinker on humanity and corporeality in today's world dominated by technology. In *Second Body* (2016), he choreographs a surreal spectacle of the transformation of a solo dancer's body from the corporeal materiality of flesh to the virtual presence of an immaterial body in a digital landscape through the visual illusion made possible by 360° Kinect motion sensing and projection. The result is simultaneously a celebration and apocalyptic warning of the taking over by the 'second body' created by digital and AI technology.

Contemporary indigeneity

One of the greatest dance traditions in Taiwan is that of the indigenous people, who have inhabited the island long before the arrival of the Han people for thousands of years. Of the now-existing 16 tribes, most of them have traditions of dance and song carried out at important events or ceremonies such as weddings, harvests, funerals, worship of ancestors, and hunting. Because of the colonial history of the island, the indigenous people's dance and song traditions suffered a long period of suppression and/or distortion due to political and sometimes religious or commercial interference. A telling example is the creation of the genre '*shandi wu*', meaning 'mountain dance', during the *minzu wudao* era. Based upon some stereotypical gestures and steps appropriated from indigenous dances, '*shandi wu*' was essentially the creation of Han society mainly for the purposes of political propaganda or tourist entertainment.

In 1991, the Formosa Aboriginal Song and Dance Troupe (Yuan Wu Zhe) was founded by a group of young indigenous people from different tribes in response to the rising Indigenous Rights Movement in the post-martial law years. Conducting extensive fieldworks for each production, the troupe produced stage performances of traditional song and dance rituals from various indigenous communities and aroused many indigenous youths' awareness of their own cultural heritage. Yet due to the unavoidable controversies over staged rituals and the legitimacy of performing ceremonial songs and dances outside their original contexts, the group took a new direction in later years by creating contemporary theatrical productions incorporating songs, dances, legends, and histories of indigenous societies. An excellent example is *Pu'ing: Tracing the Atayal Route* (2013), which ingeniously weaves historical incidents, legendary figures, and fictional narratives from the Atayal tribe to tell a complex story spanning many decades of multiple colonialisms.

Perhaps not coincidentally, Yuan Wu Zhe invited Bulareyaung Pagarlava, a former Cloud Gate dancer and indigenous choreographer of the Paiwan tribe, to direct and choreograph for *Pu'ing*, a term meaning root and route in Atayal. Forsaking all the modern dance techniques he had learned as a dancer and choreographer, Bulareyaung utilized the movement leitmotif of walking and the physical images of dancers moving solemnly in lines to convey the multi-layering of historic times and the aesthetics of iconic patterns in indigenous embroidery and woodcarving. In 2014, Bulareyaung made the important decision to move back to his hometown Taitung on the southeast coast of Taiwan and founded his own company, BDC (Bulareyaung Dance Company). Since then he has created a series of dance theatre work, astonishing the Taiwan audience, and scored the prestigious Taishing Arts Award for two consecutive years (2018, 2019), an unprecedented record for any artist in Taiwan.

Like Yuan Wu Zhe, Bulareyaung has led his dancers into various tribes to learn traditional songs and steps from elders so as to gain first-hand knowledge of their own cultural roots. However, the most important inspiration has been the lifestyle, human relationships, living environment, and contemporary issues of the indigenous society. *Stay That Way* (2017) draws upon the power of indigenous singing and alludes to corporeal imageries of oppression, vulnerability, and resilience to comment on the controversy over indigenous traditional territories, an issue triggering a long protest by indigenous artists and communities in Taipei since February 2017. *# Yes or No* (2019) borrows the formats of drag shows and jukebox karaoke singing in indigenous villages to tell the stories of the company members, who are indigenous youths with hidden wounds of dysfunctional families, past traumas of school bullying, gender identity conflicts, and so on. Transcending the boundaries between life and art, high and low, tradition and contemporaneity, BDC has expanded the scope and imaginary power of Taiwanese contemporary dance through doubly decolonizing the dancing bodies and modes of theatrical representation, from West-imported modern dance forms as well as from a concert dance tradition dominated by Han choreographers.

Taiwan, an authentic voice in dance

In December 2019, Lin Hwai-min stepped down as the artistic director of Cloud Gate after leading the company for nearly half a century, a highly symbolic moment for the dance world in Taiwan. When he choreographed *Legacy* in 1978, he asked the questions: 'Who am I? Where did I come from?' The same questions have been asked by different generations of Taiwanese choreographers and dancers since then. An island state whose status as a nation has been treated ambiguously by the global community since the 1970s and whose existence has been constantly in threat by the neighbouring superpower China, which has had complex political, cultural, and economic relations with Taiwan, those questions are often asked with particular poignancy and urgency. In the past two decades, Taiwan has trained some of the best dancers for modern, contemporary, and ballet companies in the world, testifying to the solid training systems established in the dance academies. At the same time, policies and mechanisms encouraging creativity and sponsoring international exchanges and collaborations from both the official and private sectors have played pivotal roles in nourishing young talents and supporting continuing efforts in choreography. The vibrant and diverse dance scene described in this chapter demonstrates the innovation and authenticity of the many dance voices actively engaged in dialogues with the historical, social, cultural, and aesthetic issues continuously shaping the identity and character of Taiwan as a country (CKS Cultural Center 2009).

Note: This is a revised and updated version of the original chapter in Wang, Yunyu and Burridge, Stephanie (ed.) (2012), *Identity and Diversity: Celebrating Dance in Taiwan*, Routledge, India. The second part has been thoroughly rewritten to reflect the many new developments in the past decade.

Notes

1 While the tradition of Chinese court classical dances has been lost in China, they have been kept and performed in Japan and Korea for many centuries. Liu's reconstructions are the integrated results of all her research rather than copies of the Japanese or Korean versions. In addition, her work is totally independent of the re-creation of 'Chinese classical dance' in post-1949 China. Her field trips to China were mainly to study the Dunhuang cave murals and the Dazu rock carvings. For more details, see Liu's article 'A Study of Banquet Music and Dance at the Tang Court (618–907 CE)' in *Identity and Diversity: Celebrating Dance in Taiwan* (2012).

2 Wu Kuo-chu was a theatre graduate from the Taipei National University of the Arts and later trained as a choreographer at the Folkwang Hochschule in Essen. He was appointed the artistic director of the Tanztheater, Staatstheater in Kassel, Germany, in 2004 and choreographed several outstanding pieces for German and Taiwanese dance companies. His premature death at the age of 36 in 2006 is truly a great loss to the dance world.

References

Burridge, Stephanie and Yunyu Wang, eds. (2012). *Identity and Diversity: Celebrating Dance in Taiwan.* London: Routledge.

Chang, Chung-hsuian and Yatin Lin, eds. (2005). *Lin Hwai-Min International Dance Conference Proceedings.* Taipei: Council for Cultural Affairs.

Chen, Ya-Ping. (2008). "Dancing Chinese Nationalism and Anti-Communism: *Minzu Wudao* Movement in 1950s Taiwan" in *Dance, Human Rights and Social Justice: Dignity in Motion.* Eds. Naomi Jackson & Toni Shapiro-Phim. Plymouth, UK: Scarecrow Press.

———. (2018). "In Search of Asian Modernity: Cloud Gate Dance Theatre's Body Aesthetics in the Era of Globalisation" in *Contemporary Choreography: A Critical Reader*, 2nd Edition. Eds. Jo Butterworth and Liesbeth Wildschut. London: Routledge.

———. (2020). "Exorcism and Reclamation: *Jiao* and the Corporeal History of the Taiwanese" in *Corporeal Politics: Dancing East Asia.* Eds. Katherine Mezur and Emily Wilcox. Ann Arbor: University of Michigan Press.

———. (2021). "Cheng Tsung-Lung" in *Fifty Choreographers*, 3rd Edition. Eds. Lorna Sanders and Jo Butterworth. London: Routledge.

CKS Cultural Center. (2009). *Embracing Magnificence: The Beauty of Taiwan Performing Arts.* Taipei: The CKS Cultural Center.

Liao, Chao-yang. 1995. "The Sadness of the Chinese: A Response to Chen Chao-ying on Cultural Construction and National Identity" *Chung-Wai Literary Quarterly*, 23/10. (in Chinese).

Lin, Yatin. (2018). "Choreographing a Flexible Taiwan: Cloud Gate Dance Theatre and Taiwan's Changing Identity" in *The Routledge Dance Studies Reader*, 3rd Edition. Eds. Jens Richard Giersdorf and Yutian Wong. London: Routledge.

———. (2020). "Choreographing Digital Performance in Twenty-First-Century Taiwan: *Huang Yi & KUKA*" in *Corporeal Politics: Dancing East Asia.* Eds. Katherine Mezur and Emily Wilcox. Ann Arbor: University of Michigan Press.

Lu, Yuh-jen. (2011). "Decolonized Imagination: Modernity and Modern Dance in 1970s Taiwan" *Arts Review*, 20.

16

ROOTS AND ROUTES OF CLOUD GATE DANCE THEATRE'S *NINE SONGS*

Lin Yatin

Introduction

In April 2008, I visited Ubud, Bali, in transit from a dance conference in Indonesia. Acknowledged as a main inspiration for the creation of Lin Hwai-min's full-length dance epic *Nine Songs* (1993), Ubud has even been described by Lin as his 'second home'.[1] In this chapter, I will unravel the impact of travel on contemporary artists and their creations. Through opportunities to visit other locations, either for work or leisure, I propose that such direct contact with other cultures enriches and complicates choreographers' sensitivities. I refer to issues of travel and displacement as raised by scholars such as Mary Louis Pratt, Caren Kaplan, and James Clifford, theorizing these complex questions through *Nine Songs*. After all, in our age of globalization, when travel is quite accessible and even necessary, especially in terms of stepping out and entering the international circuit of arts festivals, I offer an example of how *Nine Songs* was conceived, what it may represent, and how it has been circulated in various international dance venues.

Nine Songs, based on the Chinese work of the same title from the 3rd century BC, was developed during Lin's extensive travel within Asia between 1988 and 1991. This was during Cloud Gate's hiatus due to financial and other reasons. 1987 marked a historical moment in Taiwan. Martial law was lifted after almost half a century of banning freedom of speech, freedom of the press, and so forth. This two-and-a-half-year break allowed Lin and other company members to travel abroad and recharge.

His decision to travel to remote regions of China, Tibet, Indonesia, and India was a conscious effort to 'decolonize' himself from his previous education in the West, especially the United States. Lin wanted to study the culture of his neighbouring Asian countries, which he felt was quite foreign to him. He went to Bali and was inspired by a lotus-filled pond at Café Lotus, where outdoor performances of traditional dances were offered nightly against the spectacular facade of a Balinese temple.[2] This striking lotus pond scenario later made its way into Lin's production of *Nine Songs*.

The text *Nine Songs* by Qu Yuan was originally a collection of poems from ancient China depicting peoples' respect and fear towards the gods, the agony of lost love, and the

DOI: 10.4324/9781003160007-27

mourning of martyrs. Lin adapted this narrative as a 'springboard' for his transnational imaginary, transposing these ancient gods and goddesses into our contemporary world. More than a mere visual spectacle, it contained sharp social commentaries against tyranny throughout Chinese history.

In all, there are eight sections in *Nine Songs*, named after select deities from the poetry. The set design by renowned Chinese-American Ming-cho Lee, based on Lin's lotus motif, transfers the audience into another space and time. Throughout these sections, there are two roles which appear and reappear: namely the Shaman and the Traveler. The significance of these two roles will be mentioned later in this chapter. The following description is based on my multiple viewings of the work live and later through Cloud Gate's 2003 publication of full-length production on video.

Other than the sound of trickling water from the lotus pond located at the orchestra pit, Taiwan's indigenous songs and flying calligraphy of the *Nine Songs* text projected onto the screen opened the performance. The Traveler is the first person to appear, dressed in a black Western suit and hat, carrying a suitcase. In contrast, the rest of the dancers are dressed in loose white robes, as if participating in a ritual. Further enhancing the lotus motif, the proscenium stage is framed with movable panels of enlarged lotus drawings, referring to the Buddhist life cycle of birth and rebirth.[3]

The female Shaman in the scene 'Greeting the Gods' steps out dressed in a bold red dress. She undulates her body, shaking every joint and muscle at the same time. Displaying her deeply inscribed Graham contraction technique, she resembles a Taiwanese shaman gone into trance. The Sun God is then introduced through the lifted back panel. Standing tall, he is balanced on the shoulders of his two attendants. Soft Tibetan bells usher in a sense of peace, contrasting sharply with the confrontational duet between the female Shaman and the male God.

Next, looming voices of Tibetan monks represent the oppressive atmosphere in the section 'Gods of Fate'. The movement vocabulary, derived from contact improvisation, expresses male dancers' harsh manipulation of female dancers. Lin achieved such effects by transforming a formerly 'meaning-neutral' exercise called 'Molding Your Partner' – where one person passively accepts the instructions given by the partner's push or tug. Here, the Traveler reappears not only once but at the beginning and towards the end of this section: riding across the stage in a bike, then carrying an umbrella and a suitcase, as if an onlooker observing a scene from Michelangelo's *The Last Judgement* from an angle above, emotionally detached.[4]

The following sections are expressed through movement vocabularies derived from Javanese court dance, Chinese operatic movement, and *butoh*-like aesthetics, respectively. The Goddess of the River Xiang employs the curvilinear Javanese court dance to Taiwan's indigenous songs and slow gamelan music, expressing the introverted, inner world of a lonely maiden. Eventually, the Shaman walks towards the Goddess and rips off her mask. Not long after, the Traveler also reappears, walking across the stage behind the saddened Goddess without any interaction whatsoever.

In the later section 'Homage to the Fallen', the ensemble in white robes re-gathers on stage, wearing white headbands, symbolizing mourning. The names of martyrs – ranging from those in early Chinese history up to present-day Taiwan – are announced randomly in the dialects of the fallen heroes. As the drumming accelerates and becomes louder, suggesting the shooting of machine guns, anonymous victims lined up in a row drop to the floor one by one, reminiscent of the execution scenes from the 2–28 Incident from 28 February 1947, that resulted in uncalculated deaths among the Taiwanese elite caused by the KouMingTang

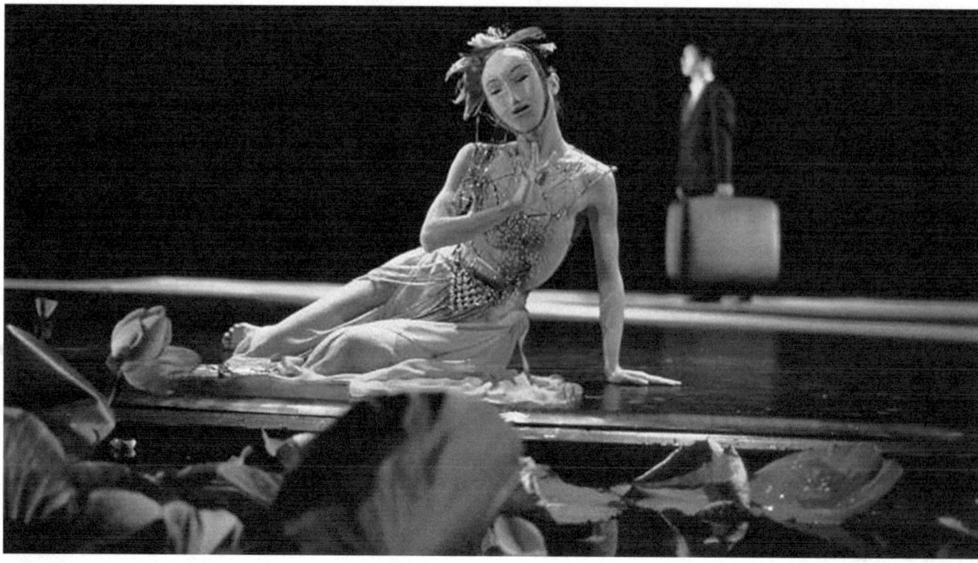

FIGURE 16.1 *Nine Songs*

Source: Performed by Cloud Gate Dance Theatre of Taiwan

Dancer: YANG Mei-jung

Photo by LIU Chen-hsiang (courtesy of Cloud Gate Dance Theatre of Taiwan)

(KMT) Chinese Nationalist Party from mainland China. Two parallel rays of light shine towards the audience like headlights from a vehicle. A man blocking the lights, arm stretched out in protest, clearly evokes the image of the Chinese youth in the Tiananmen Incident who defiantly stood in the path of the approaching military tank. As if he is 'struck' by an invisible bullet and about to hit the ground, the Shaman rushes in to keep him from falling. She carries him to the edge of the lotus pond and cleanses him. Unlike the Traveler, who reappears in his black suit and luggage, detached, the Shaman throws herself into each scenario, dedicating her body and soul in the hope of bringing comfort to her people.

Through the roles of the female Shaman and the male Traveler, audiences were guided along this journey from past to present, East and West, as well as from the realm of the deities to the mortals. In the end, Lin provided solace for those who witnessed and/or sacrificed for their nations. Two perspectives resolve that are beyond our power – we can either choose to be totally committed, like in the motherly quality of the Shaman, or stay collected and observe from a distance, like the Traveler.

Travel and transculturation in *Nine Songs*

Prior to the premiere of *Nine Songs*, Lin admitted that this dance is a reflection from his travels.[5] Chang Chao-tang – photographer, documentary director, and longtime collaborator of Lin Hwai-min – claimed that the Traveler is the main protagonist who threads the whole dance together. He wrote:

> the appearance of the traveler is a response as well as a search; seemingly irrelevant, yet [totally] captivating, he guides the audience to another place [and time].[6]

Lin also confessed that the Traveler in this dance represented himself, reflecting upon what he saw, heard, and thought about throughout his two to three years of travelling.[7] Like a character from Magritte's painting, the Traveler fits into our image of the typical 'Western individual, usually male, "white," of independent means, an introspective observer, literate, acquainted with ideas of the arts and culture, and, above all, a humanist'.[8]

A former bestselling author of short fiction, Lin may be continuing the legacy of 'travel record literature' from China through dance, especially in the style of famous poet Su Shih, whose travel writing presented 'a philosophical and moral argument as its central purpose'.[9] Scholar Chan YinHa (2005) also discussed the issue of travel in the tradition of the Chinese literature, but in regard to the postcolonial context of Hong Kong.

As for travelogue-choreographies, this genre has also gained currency in the age of globalization. Pina Bausch and her commissioned co-productions between Germany and international cosmopolitan cities are famous examples. In addition, African-American choreographer Ralph Lemon's *Geography Trilogy* and Chinese-American choreographer Shen Wei's trilogy simply titled *Re-1, Re-2*, and *Re-3* are all heavily based on research from extended periods of travels.[10]

Mary Louis Pratt, in her book *Imperial Eyes: Travel Writing and Transculturation*, wrote that 'Ethnographers have used this term [transculturation] to describe how subordinated or marginal groups select and invent from materials transmitted to them by a dominant or metropolitan culture'.[11] This concept of transculturation is evident in Lin's adaptation of Western dance techniques such as incorporating Graham and contact improvisation into *Nine Songs*. On the other hand, Lin also borrowed from other Asian cultures, such as gamelan music and the fluttering hand gestures from Javanese court dance.[12] Interestingly, Lin's Indonesian friend, dance scholar Sal Murgiyanto, did not see this section as 'Indonesian'. He felt Lin intelligently merged the resemblance between the calm energy of Javanese dance with the sense of flow from *tai chi* and other Chinese movement traditions.[13]

Nevertheless, scholar Chen Kuan-hsing from Taiwan alerted against the perils of Taiwan's possible 'imperialistic' ideology of 'Marching South', that is, in relation to its neighbouring Southeast Asian countries. Proposed by Taiwan's former President Lee Teng-hui in the mid-1990s, the close affinity of this discourse to the 'Greater Asia Co-Prosperity Sphere' concept from the Japanese Occupation period of Taiwan is quite alarming. As Chen argued, it is based on the convergence of three axes: nation-building, state-(re)making, and (sub)empire-forming, and thus deserves our attention.[14]

Touring routes of Cloud Gate: dance festivals as contact zones and beyond?

Feminist experimental ethnographies have interrogated the history of travel writings and are critical of their imperialist and often male-dominated legacy, not to mention the issue of class in allowing the possibility for leisurely travel.[15] James Clifford, in his study on travel and translation, applied Pratt's concept of 'contact zone' to his own case study of museums.[16] According to Pratt, 'contact zone' refers to:

> space of colonial encounters . . . in which peoples geographically and historically separated come into contact with each other and establish ongoing relations, usually involving conditions of coercion, radical inequality, and intractable conflict.[17]

She further explains:

> I . . . foreground the *interactive, improvisational* dimensions of colonial encounters.
> A 'contact' perspective emphasizes . . . *co-presence, interaction, interlocking understandings and practices.*
>
> *(author emphasis)*[18]

Drawing upon my multi-locale fieldwork of travelling with Cloud Gate on their tours to various dance festivals in Europe and the United States, as well as Australia and New Zealand during the year 2000, I propose the notion of dance festivals as contact zones where companies from across the world perform on the same stage. Evidently, various economic and political issues come into play, including sources of funding, ideologies behind these funding policies, and so forth, referred to collectively here as the 'political economy of touring'.[19] I explore how these festivals achieve visibility for dance companies from the non-Western world, provide opportunities for interaction with artists and audiences from different countries, and allow for negotiations of the artists' space and place.

Sydney Olympic Arts Festival

The turn of the millennium coincided with optimism in the world economy. Perhaps the most publicized foreign tour by Cloud Gate in 2000 was their appearance at the Sydney Olympics Arts Festival, such as covered by Cherie Huang (2000) in Taiwan's *Performing Arts Review (PAR)* monthly from the National Theatre. During the Olympics, athletes from various nations compete for medals to win honour for their countries. This display of patriotism involves the display of national anthems, national flags, and so forth. However, Taiwan, or 'Republic of China', as it is officially called, is only allowed to participate in the Olympics under a special condition. According to the People's Republic of China led by the Chinese Communist Party, Taiwan is a renegade province. Thus, Taiwan's athletes are only allowed to compete in the Olympics under the banner of 'Chinese Taipei'. In other words, China obstructs Taiwan's admission into international organizations (including the UN, the WHO, and even arts festivals). Therefore, during the Sydney Olympics Art Festival, for Cloud Gate to be chosen as the sole representative from Asia – on a par with Bill T. Jones from the United States, Pina Bausch from Europe, and hometown favourite DV8 Physical Theatre (its artistic director Lloyd Newson is a native of Australia) – was a significant event.

As Benedict Anderson argued, newspapers enhance our sense of national identity.[20] With the current popularity of other media (such as television and the internet), Cloud Gate's Australian tour was fully documented and publicized by Taiwan's Public Television Service (PTS). To cover all these related expenses, Cloud Gate attracted extra donations from private sectors – especially high-tech companies from Taiwan such as laptop computer designer and manufacturer Acer.

Lyon Dance Biennale 'Silk Road'

On the other hand, the festival scene at Lyon Dance Biennale, an ongoing festival with its own reputation, presented a totally different phenomenon. For 'Silk Road,' Artistic

Director Guy Darmet travelled extensively along this historic route from Asia to Europe to select the companies himself. As a result, out of the 32 groups invited, 5 came from Taiwan: Cloud Gate Dance Theatre, Legend Lin Dance Company, HanTang Music and Dance Ensemble, U-Theatre, and the Taiwan National Theatre Academy, whose acrobatic students paraded down the main streets of Lyon as part of their outdoor attractions.

However, these representatives from Taiwan were not the only groups from a Chinese heritage. Hong Kong's premiere dance company, the City Contemporary Dance Company (CCDC), presented Helen Lai's own production of *Nine Songs* (1991) to Chinese composer Tan Dun's contemporary opera music, while China also sent newly formed modern dance groups from Beijing and Shanghai.[21] Yet due to China's political protest, the Lyon Dance Biennale innovatively categorized each dance company by its city of origin (such as Taipei) instead of the tradition of using names of countries, thus allowing Taiwan's participation in this important international event.

Brooklyn Academy of Music

According to Lin Hwai-min and other choreographers from Taiwan among his peers, New York has been viewed as a Mecca for the arts. For Cloud Gate to perform at the 'center' of concert dance in the West signifies the 'arrival' of Lin and his dancers in the international global dance scene. Cloud Gate's performances at the Next Wave Festival of the Brooklyn Academy of Music (BAM) confirmed its status as a mainstream performing arts company, on a par with the other 'A-list' BAM-invited performers.

Munich Dance Festival

Referring back to Clifford's study on museums as contact zones which emphasize the aspect of reciprocity, I also witnessed a parallel example between Lin Hwai-min and renowned Indonesian choreographer Sardono Kusumo in the year 2000. Both artists were invited to the Munich Dance Festival and Forum – 'Between the Worlds', a unique format which invited artists to dialogue through the three-day Choreographers' Forum.[22] Lin, after seeing Sardono's unique use of voice and body movement in a performance, invited him to give a special master-class for Cloud Gate dancers in Munich.[23] These are just some examples of interaction and improvisation between artists from different countries in a festival 'contact zone' setting. More recently, young Indonesian male choreographer and dancer Danang Pamungkas joined Cloud Gate 2, performing in Lin Hwai-min's choreography *Song of the Birds* (2008).[24] It is due to such reciprocity between the 'traveler' and the host country that long-lasting relationships can be fostered.

Epilogue: 'The Wanderer's Project' – Lin's travel initiative for young people to explore Asia

In 2003, Lin Hwai-min was awarded the National Culture Award by the government in Taiwan. He generously donated his award money towards his new initiative, 'The Wanderer's Project', to provide travel grants for promising youths wishing to explore Asia.[25] Lin published a book of touring with Cloud Gate in 2007 across eight cities in Europe over seven weeks. Titled *Wandering with Cloud Gate*, this travelogue includes accounts of the joys and wonders of travelling with his beloved dancers and crew while inserting the pages with thoughts of family, friends, and events back home.[26]

I have outlined various possibilities travel can enrich the lives and work of artists and scholars. Through this whirlwind tour of the root and routes of Lin Hwai-min's *Nine Songs*, I hope to raise issues worth contemplating in terms of travelling bodies and their impact on dance-making, identity, and hybrid cultures.

Note: This chapter was originally published in Wang, Yunyu and Burridge, Stephanie (ed.) (2012), *Identity and Diversity: Celebrating Dance in Taiwan*, Routledge, India.

Notes

1 Personal communication with Lin Hwai-min, Taipei, 29 May 2008.
2 I attended one such performance at Lotus Café in Ubud on 3 April 2008. The programme consisted of Balinese Barong dance, Kris dance, gamelan musical concert, and others.
3 See Lin Hwai-min, Hsu Kai-chen, and Chi Hui-ling, [*On the Making of Nine Songs*] (Taipei: *Min-sheng Daily*, 1993), pp. 18–26. The lotus paintings are reproduced with permission from Taiwanese painter Lin Yu-shan.
4 Lin Hwai-min, [*Material and Transformation: A Choreographic Journey of the Dance Drama Nine Songs*] (Taipei: Cloud Gate Dance Foundation, 1995), pp. 63–64.
5 Lin Hwai-min, "Song of Prayers: Written before the Premiere of *Nine Songs*," *United Daily News*, 8/8/1993.
6 Chang Chao-tang, "The Traveler *Nine Songs*," *China Times*, 9/4/1993.
7 Lin Hwai-min, [*Material and Transformation: A Choreographic Journey of the Dance Drama Nine Songs*], p. 23.
8 Caren Kaplan, *Questions of Travel: Postmodern Discourses of Displacement* (Durham, NC: Duke University Press, 1996), p. 50.
9 See http://en.wikipedia.org/wiki/Travel_literature, accessed March 19, 2008.
10 See map listing Pina Bausch's 14 city-commissioned works from *Pina Bausch's Dancing for the World*, edited and co-authored by Yatin Lin (Taipei: National Chiang Kai-shek Cultural Centre, 2007), pp. 96–97.
Ralph Lemon's *Trilogy*, which is well documented with a DVD set (2007) and writings (2000, 2004) by the choreographer himself, consists of: *Geography* (1997), *Tree* (2000) and *Come home Charley Patton* (2004).
Shen Wei's trilogy is based on his travels to Tibet, Angkor, and the Silk Road, respectively (2006, 2007, and 2009). Telephone interview with Shen Wei (from Taipei to Beijing), 5 July 2008.
11 Mary Louis Pratt, *Imperial Eyes: Travel Writing and Transculturation* (London and New York: Routledge, 1992), p. 6.
12 Indonesian dancer Ms. Endang Nrangwesti, wife of Sal Murgiyanto, taught Javanese court dance at Cloud Gate in 1993, and has continued teaching Indonesian dance at Taipei National University of the Arts (TNUA) since then.
13 Dr. Sal Murgiyanto, Personal interview. 7 March 2008. Taipei: TNUA.
14 Chen Kuan-Hsing, "The Imperialist Eye: The Cultural Imaginary of a Sub-Empire and a Nation-State" (in Chinese), *Taiwan: A Radical Quarterly in Social Studies*, No. 17, July 1994, pp. 149–222. English version translated by Wang Yiman, *Positions: East Asia Cultures Critique*. Volume 8, Number 1, Spring 2000, pp. 9–76.
15 See Janet Wolff, "On the Road Again: Metaphors of Travel in Cultural Criticism," in *Cultural Studies* 1993, 7(2), pp. 224–239.
16 James Clifford, "Museums as Contact Zones," in *Routes: Travel and Translation in the Late Twentieth Century* (Cambridge, MA and London, England: Harvard University Press, 1997), pp. 188–219.
17 Pratt, *Imperial Eyes: Travel Writing and Transculturation*, p. 6.
18 Ibid, p. 7.
19 Based on my earlier writing on Cloud Gate, I'm borrowing from Marta Savigliano's study of *Tango and the Political Economy of Passion* to coin the phrase "political economy of touring." See my Chinese-English bilingual article from 2001 (Lin 2001).

20 See Benedict Anderson's *Imagined Communities: Reflections on the Origin and Spread of Nationalism* (London and New York: Verso, 1991).

21 Choreographed before Lin Hwai-min's version, Helen Lai's *Nine Songs* (1991) started out as a collaboration with Chinese composer Tan Dun for the annual Hong Kong Arts Festival in the form of an experimental opera involving both singers and dancers. Accompanied by a live orchestra, with this version of *Nine Songs* is less literal than Lin Hwai-min's interpretation, using the central element of different-colored sand to mark the changes in the different landscapes of the work. For more coverage on the Lyon Dance Biennale, see my article in Chinese from the *PAR* monthly in 2001.

22 Based on my article on the Munich Dance Festival in Chinese (see Lin 2000), themes for the Choreographer's Forum included 'Thinking from the Body,' 'Tradition and Modernity,' and 'Local/Global.' Dance scholars Andre Lepecki and Ghislaine Boddington co-moderated, guiding the audiences, including myself, through various thought-provoking sessions. Lepecki also wrote an article on Lin's *Nine Songs* based on its tour to BAM in 1996.

23 This was an improvised decision by Lin Hwai-min, laying the grounds for future collaborations – including inviting Sardono to the Asia-Pacific Arts Forum held at the Taipei National University of the Arts later that summer (2000).

24 Although Danang has since left Cloud Gate 2 and become a choreographer in Indonesia, he has collaborated with other Taiwanese dance artists such as Su Wen-chi of YiLab in her works *Infinity Minus One* (2018) and *Anthropic Shadow* (2019). See Su Wen-chi's website: www.suwenchi.com/infinity-minus-one.

25 In The Wanderer's Project website, Lin Hwai-min recalls his fond memories from backpacking across Europe before returning to Taiwan, having just finished his graduate studies in the United States, and the impact it had on his career. Thus Lin donated his NT$1,000,000 (equivalent to approximately US$33,000) award as seed money for this travel grant. As of 2020, the grant is still being handed out, thanks to donations from the private sector. See http://site.cloudgate.org.tw/wanderer/about.html (accessed 26 May, 2020). Among the many recipients, Hsieh Wang-lin (2008) published his book *Circling the Mountain* upon returning from his travels with this grant.

26 Lin likens them to *nomads*, shifting from locale to locale across a brief period of time, never staying long enough in one place.

References

English

Anderson, Benedict. *Imagined Communities: Reflections on the Origin and Spread of Nationalism*. London and New York: Verso, 1991.

Chen, Kuan-Hsing. (English translation by Wang, Yiman). "The Imperialist Eye: The Cultural Imaginary of a Sub-Empire and a Nation-State," in *Positions: East Asia Cultures Critique*. Volume 8, Number 1, Spring 2000, pp. 9–76. (Original Chinese version published in 1994, see reference in Chinese).

Clifford, James. *Routes: Travel and Translation in the Late Twentieth Century*. Cambridge, MA and London, England: Harvard University Press, 1997.

Kaplan, Caren. *Questions of Travel: Postmodern Discourses of Displacement*. Durham, NC and London: Duke University Press, 1996.

Lemon, Ralph. *Geography: Art/Race/Exile*. Middletown, CT: Wesleyan University Press, 2000.

———. *Tree: Belief/Culture/Balance*. Middletown, CT: Wesleyan University Press, 2004.

Lepecki, Andre. "Postcolonialism, Interculturalism," in *Ballet International/Tanz Aktuell* (January 1996). Berlin: Friedrich Berlin Verlag, pp. 36–41.

Lin, Yatin. "Dancing in the Age of Globalization: Cloud Gate Dance Theatre and the Political Economy of Touring," in *Dance Studies and Taiwan: The Prospect of a New Generation*. Ed. By Chen Yaping and Chao Chifang. Taipei: National Chiang Kai-shek Cultural Center, 2001. pp. 200–212.

Pratt, Mary Louis. *Imperial Eyes: Travel Writing and Transculturation*. London and New York: Routledge, 1992.

Savigliano, Marta E. *Tango and the Political Economy of Passion*. Boulder, CO and Oxford: Westview Press, 1995.

Wolff, Janet. "On the Road Again: Metaphors of Travel in Cultural Criticism," in *Cultural Studies*. Volume 7, Number 2, 1993, pp. 224–239.

Chinese

Chang, Chao-tang 張照堂. "The Traveler *Nine Songs* 〈旅人《九歌》〉," *China Times*, September 4, 1993.
Chen, Kuan-hsing 陳光興. "The Imperialist Eye: The Cultural Imaginary of a Sub-Empire and a Nation-State 〈帝國之眼: "次"帝國與國族 – 國家的文化想像〉," *Taiwan: A Radical Quarterly in Social Studies*《台灣社會研究季刊》, No. 17. July 1994, pp. 149–222.
Chan, Yin Ha. 陳燕遐. "[Why Travel? The Metaphor of Travel in Postmodern/Postcolonial Cultural Discourse] 〈旅行, 所為何事? – 後現代/後殖民文化論述中的旅行隱喻〉," conference paper from 2005 Cultural Studies Conference in Taiwan (去國 汶化 華文祭2005 年華文文化研究會議).
Hsieh, Wang-lin 謝旺霖. [*Circling the Mountain*.] 《轉山 – 邊境流浪者》. Taipei: Yuanliu Publishing, 2008.
Huang, Cherie. 黃琇瑜. "[Report on the Sydney Olympic Art Festival.] 〈人我裡外, 各抒心境: 四大舞團在奧運藝術節相映爭輝〉." *PAR Performing Arts Review*, No. 95. November 2000, pp. 15–17.
Lin, Hwai-min 林懷民. [*Material and Transformation: A Choreographic Journey of the Dance Drama Nine Songs*]. 《素材與變貌: 舞劇九歌的編作歷程. Taipei: Cloud Gate Dance Foundation, 1995》.
———. "Song of Prayers: Written before the Premiere of *Nine Songs*," 〈禱歌 – 寫在《九歌》首演前〉 in the *United Daily News*, August 8, 1993.
———. [*Wandering with Cloud Gate: A Diary of Traveling Eight Cities in Europe over Seven Weeks*.] 《跟著雲門去流浪:七週八城的歐洲巡演日記》. Taipei: Locus Publishing, 2007.
Lin, Hwai-min, Hsu Kai-chen 徐開塵, and Chi Hui-ling 紀慧玲. [On the Making of *Nine Songs*.] 《喧蟬鬧荷說九歌》. Taipei: *Minsheng Daily*, August 1993.
Lin, Yatin 林亞婷, editor and co-author. 《碧娜·鮑許為世界起舞》 [*Pina Baush: Dancing for the World*]. Taipei: Chiang Kai-shek Cultural Center, 2007.
———. "[Report on Lyon Dance Biennale] 台灣舞團集體躍上國際舞台 – 里昂雙年舞蹈節絲路之旅," *PAR*, No. 97. January 2001, pp. 7–11.
———. "[Report on Munich Dance Festival] 搭起世界的橋樑? 記慕尼黑國際舞蹈節," *PAR*, No. 91. July 2000, pp. 28–32.

Interviews

Lin, Hwai-min 林懷民. Taipei, May 29, 2008.
Murgiyanto, Sal. Personal interview. Taipei National Uiversity of the Arts, March 7, 2008.
Shen, Wei 沉偉. Telephone interview. (from Taipei to Beijing.) July 5, 2008.

Videography

Cloud Gate Dance Theatre 雲門舞集 of Taiwan. *Nine Songs* 九歌. Choreographed by Lin Hwai-min. Filmed by Chang Chao-tang 張照堂. Taipei, Jingo: 2003.
Ralph Lemon. *Geography: Trilogy* DVD. New York: Mapp International Productions, 2007.

Websites

Cloud Gate Dance Theatre of Taiwan. "The Wanderer's Project," Website. http://site.cloudgate.org.tw/wanderer/about.html, accessed 26 May, 2020.
Su Wen-chi and YiLab: www.suwenchi.com/infinity-minus-one, accessed 25 May, 2020.
"Travel Literature," Wikipedia entry. https://en.wikipedia.org/wiki/Travel_literature, accessed March 19, 2008.

Thailand

17

REINVENTING HOW WE MOVE

The Pichet Klunchun Dance Company in the context of contemporary Thai dance

Pornrat Damrhung and Lowell Skar

Introduction

Pichet Klunchun is a well-known dancer and choreographer from Thailand. Since directly training in Thai classical mask dance (Khon) with renowned master of the 'Giant' role Chaiyot Khummane and receiving his bachelor's degree in Thai classical dance from Chulalongkorn University more than a quarter-century ago, he has sought to embody the spirit of Thai dance on the global dance scene. His diverse work over more than 25 years has made him a key figure of Thai dance and made him well known through his work abroad, which started in 1999. Over this period, his work has evolved considerably, partly from serendipity and partly from design, but always based on disciplined work, often by working through new possibilities for dance discovered in the Khon tradition. Since founding the Pichet Klunchun Dance Company in 2007, his vision for Thai dance has expanded and diversified through a talented group of professional performers who seek to bring to life a shared experience of Thai dance today. This chapter will examine two pivotal dances that the Company developed and performed between 2012 and 2017 *Tam Kai* ('Follow the Chicken') and *Dancing with Death* – each working on the edges of the Khon tradition, and consider how these works helped to given the Company its current focus and direction. Together, these two pieces sought to break with the earlier elegant pieces more rooted in the movements and conventions of traditional Thai classical dance that were a hallmark of the company and to seek, instead, new ways to practice freedom. The chapter will conclude with a look at the larger ecology of contemporary Thai dance and locate the Pichet Klunchun Dance Company in it.

More than ten years ago, the Company developed the piece *Black & White*, the first Thai work commissioned by the Esplanade Theatres on the Bay in Singapore.[1] The troupe spent a month in Singapore as artists-in-residence putting the final touches to the piece for its premier at the Esplanade as part of its da:ns festival 2011.[2] It was a contemporary work based on Khon, classical Thai masked dance, which was visually black and white and focused on showing the importance of balance in dance – both the balance of the dancers' bodies and the balance of moral values. Since that time, this piece has toured to the Netherlands,

DOI: 10.4324/9781003160007-29

Switzerland, Japan, and many other places, including in Thailand in the fall of 2019. By most accounts, it was a great success. But Pichet saw it as leading the Company into an aesthetic dead end, where the Company could competently perform a recognizable form of beautiful contemporary dance from Thailand. To keep moving, he needed the Company to break from its choreographic cul-de-sac, but he did not know how to do it. The big question from 2013 became: How could the Company change? How to 'destroy the beautiful form of the dancers?'[3]

Creating a new environment to work through the company's problem

Fresh off the successful staging of *Black & White* in Asia and Europe, the Pichet Klunchun Dance Company hesitated at its next step. It could follow the path opened up by this and similar pieces it had done in the early 21st century, assured it could continue to appeal to diverse global audiences. The Company had developed a distinctive way to rework, repackage, and *evolve* pieces from Khon – classical masked Thai dance – for the contemporary stage in ways that could challenge both dancers and audiences enough yet remain interesting to both. While in Khon performances, dancers embody the characters and stories of the *Ramakien* (the Thai version of the *Ramayana*), which gives the piece its meaning, dancers in *Black & White* performed as contemporary dancers outside the context of the *Ramakien*, not as Khon characters, so their movements needed to give the piece meaning. Being asked to perform outside the Khon roles they had trained their lives to be challenged the dancers, since they were trained to become a specific character in their performance, and they often worried about doing the 'wrong' thing for the character they learned to embody. To deal with this problem, in *Black & White*, the dancers were guided by Pichet's strict choreography, along with the soundscape of the critically acclaimed *guqin* player Wu Na and the lighting design by Miura Asako, which provided a new environment for the dancers. They worked with Pichet to develop new Thai dance poses, movements, and interactions which differed from the classical ones they had learned as second nature in the characters from the *Ramakien*. Through his work, the Company developed a new corporeal language for Thai dance derived from Khon. By reworking the original meanings of Khon's physical performance language through Pichet's new choreography, the dancers' movements bear only a modern family resemblance to those of classical Khon and embody very different meanings and nuances than those of the *Ramakien* and Khon. The piece showed one way to make contemporary Thai dance by reworking Khon. In seeking to open up dancers' bodies and minds, *Black & White* staged a modern embodiment of the spirit of traditional Khon. Such reworking is forbidden in the framework of the Thai tradition and even offensive to many classical Thai arts, but these radical features of the work are likely missed by most foreign audiences.

If the Company would have continued on this path, it would have helped to portray Thai dance in a new way, but it would have been new Thai dance. So, if the Company wanted to do more to make new dance in Thailand, how to do it? Unsure of how to move ahead after *Black & White*, the Company first tried to solve this problem through experimentation, trying new exercises and developing different ways for its dancers to be new. For the Company, the problem of doing new dance was the problem of first changing the bodies and the minds of the dancers.

Solving the problem of encouraging dancers to free up their minds involved creating a new condition and environment for them to work in. While moving to new physical locations was possible to produce this new working environment, it was expensive and time

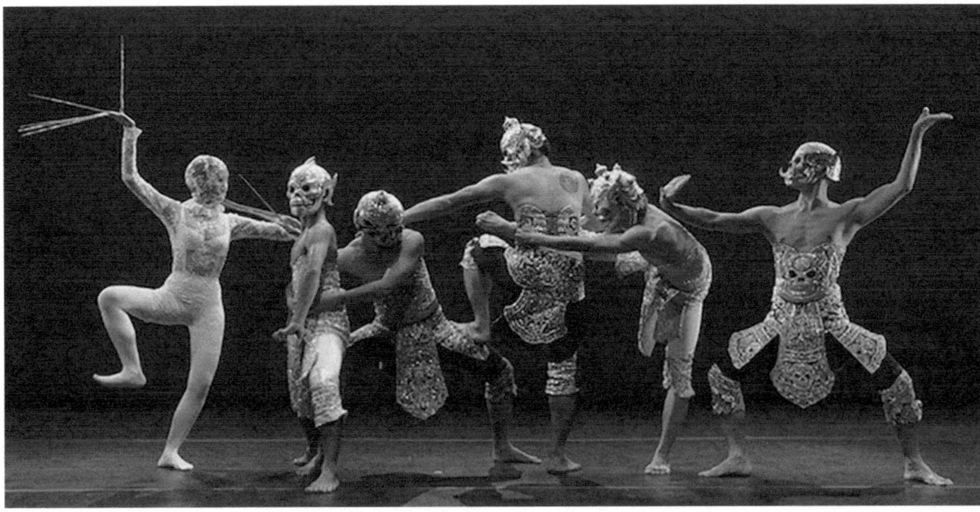

FIGURE 17.1 *Black & White* (2011)

Source: Pichet Klunchun Dance Company

Photo: Pichet Klunchun personal collection

consuming. Moreover, since the biggest problem was how to rework the dancers' minds, creating a new environment and condition for their work could be done by altering the mental space of studio where they worked. Since the dance environment the Company uses always involves sound, the Company tried to change the working sounds for its dance experiments. Testing a variety of soundscapes for the Company to experiment with eventually led to the discovery of the 45-minute-long recording of the *Kraw Nai Tang Tawada* played by the renowned master of the *ranad ek* (alto xylophone) Boonyoung Katkong. This work consists of two classical tunes often played during a theatrical performance or a ceremony, and it is normally performed by a master to show the blend of his skill, experience, intelligence, and creativity. Although not used for dance, the rapid, complicated music is useful to challenge dancers to move in creative ways.

The complex rhythms, intense pace, and sheer length of the *Kraw Nai* piece provided the Company's dancers with a dynamic new environment for them to work through their problem of what was next for them. The experimental exercises for the Company's dancers first sought to discover if the dancers could synchronize their movements with the music. After showing they could do that, the dancers were next challenged to interconnect and to communicate with one another through movements and to corporeally reflect on situations or stories while performing them in real time with the music. This experimental environment proved aesthetically productive.

Exercises in freedom: *Tam Kai* (Follow the Chicken)

The experimental studio exercises aiming to let dancers individually and collectively free up their dance movement developed over two years, from 2011 to 2013. Using the sonic space of Master Boonyoung's *Kraw Nai* performance for their new environment helped the Company to work toward a new performance, which became *Tam Kai*. Like the studio exercises of freedom at its source, the performance required the dancers themselves to do both the choreography and to dance it on the spot, live in the middle of the music, with no costumes and on a bare stage, in relation to other dancers and any audience that might be watching. For the Company, the goal was to create a space where their bodies, thoughts, and spirits were free in the here and now of the music. Coinciding with a long period of political turmoil in Thailand, the Company's exercises in freedom resonated with the ongoing struggles for a democratic society, too.[4]

Those who have seen *Tam Kai* will not easily connect it to the earlier work of the Company. It represents a deliberate break with the past. Derived from exercises practiced by the Company, it is 'a dance production that doesn't look like a dance production,' as Pichet Klunchun said.[5] It was first performed in late January 2013 at Chulalongkorn University as part of the 'Our Roots Right Now' research forum and festival on Thai and ASEAN dance and theatre, supported by the Thailand Research Fund. The deeper source of the title come from Phra Lor, the legendary hero from the literary work *Lilit Phra Lor*, who was lured by a magical 'magnificent rooster' to go to another city, where he fell in love with two princesses, which led to their doom. Pichet uses *Tam Kai* to question whether contemporary Thai audiences are being lured by the beauty of classical Thai dance and so neglecting its value beyond just glittering costumes and elegant movements. By emphasizing individual and collective improvisation, the piece also asks whether rigid Thai dance traditions restrict the ability of dancers to contribute creativity and freedom to the contemporary dance world. Viewers of the hour-long piece also may wonder about the relation of Thailand's contemporary dance not only to its past but also to the dance scenes outside of Thailand, which have long been open to foreign cultural influence.

In the *Tam Kai* performance, done for the duration of the *Kraw Nai* performance, the dancers, as the magnificent fowl underlying the title of the dance, have been encouraged to 'follow', in modified and distinct ways, the music and tune into the corporeal vocabulary of the dancer preceding them in order to move anew, to practice freedom, while being guided by an evolving improvisational sense like in the accompanying music. This piece emerges in an environment where performers need to leave behind their traditional training, forms, and feelings so they can spontaneously embody a sense of free movement with others doing the same. Viewers watch the dancers imaginatively respond not only to the driving music of *Kraw Nai* but also showing the dancers perform solos in succession, seeking to connect their movements to their fellow dancers. What Pichet aimed for was for 'the music to challenge the dancer, demanding they creatively respond to the music's powerful and fast-paced rhythms. I wanted the dancers to work with their bodies as much as they could and to check their movement with music'.[6]

Through its lengthy development in 2011 and 2012, *Tam Kai* eventually evolved into a piece with six interconnected parts, plus a concluding short collective dance in homage to its Phra Lor source. The six different parts sought to challenge different aspects of the Company's dancers' abilities to move freely and creatively, both on their own and with others, in a limited spatial and musical environment: 1) to be in tune with oneself, 2) to be in synch with the person dancing before you, 3) to be connected to the music, 4) to be part of the performance space, 5) to connect to oneself or the group, and 6) to connect to the story of Phra Lor following the rooster. The final collective dance performed after the *Kraw Nai* piece ends is performed to a different elegant piece of music and provided the underlying rationale for the dances in the previous six parts. Performed in a blend of traditional and contemporary styles, it reminded viewers of the dignified power of classical

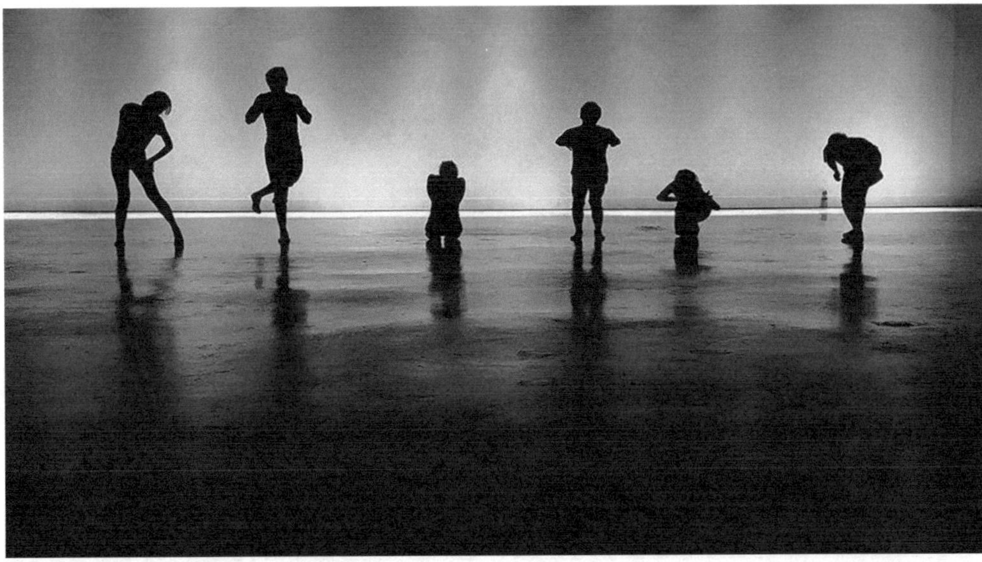

FIGURE 17.2 *Tam Kai* (2013)

Source: Pichet Klunchun Dance Company

Photo: Pichet Klunchun personal collection

Thai dance in the contemporary world. The order of the six parts and the order in which of the Company's dancers began or followed the others changed from performance to performance, ensuring that spontaneity and improvisation would remain the focus for the dancers each time the piece was performed. What began as an experimental exercise for Company members' practice in generating confidence in their own spontaneous reactions to the music and the space of performance and trust in interacting with their fellow dancers ended in clearing a new space for the troupe to exercise freedom. As such, it retained the live, experimental, and open-ended feel in each performance, too.

After performing *Tam Kai* at Chulalongkorn University in early 2013, the Company was invited to perform the piece in Kuala Lumpur, seven cities in the Netherlands, Japan, Singapore, and Hong Kong and several times in Bangkok, including with Bruce Gaston – a student of Master Boonyoung – performing live on the xylophone. Since the performance began as an exercise to free up movement in dance, the Company found it was useful to include this type of activity in workshops in places where it performed the piece, even to include some workshop participants in some of the performance. This has been an important way for the Company to extend its activities to bigger and more engaged groups of people interested in working with new ways of exercising body and mind while touring.

If *Tam Kai* was useful in 'destroying the beautiful form of the dancers' in his Company, then what to do with the dancers' broken forms? What was next for the Company? Were there better ways to work creatively and take more steps away from their training and tradition toward freedom? Those were the questions which challenged the Company in the next phase of its development. Answering those questions took the Company far from Bangkok and down a distant, dark, but lively, road that included *dancing with death*, the name of its next project that focused on putting free movement to work.

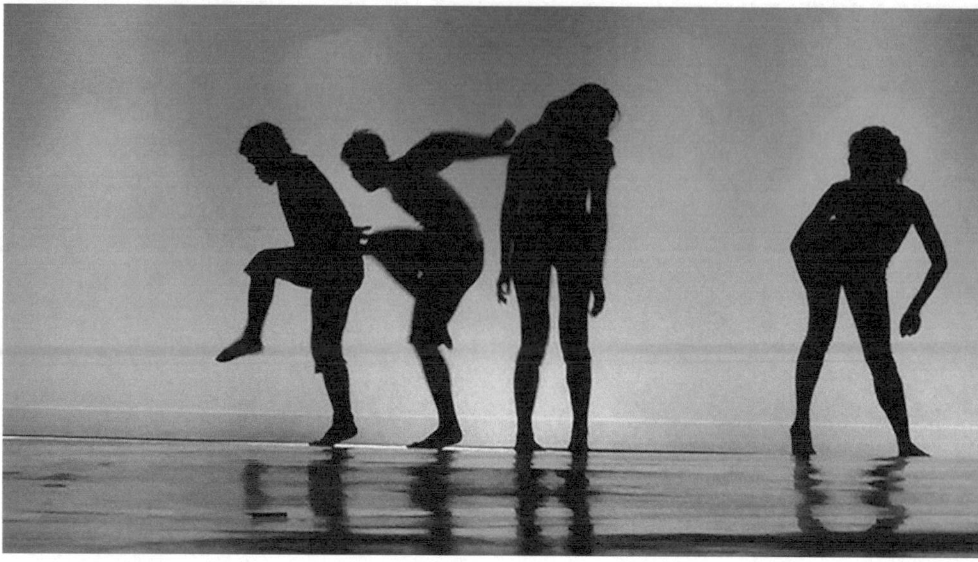

FIGURE 17.3 *Tam Kai* (2013)

Source: Pichet Klunchun Dance Company

Photo: Pichet Klunchun personal collection

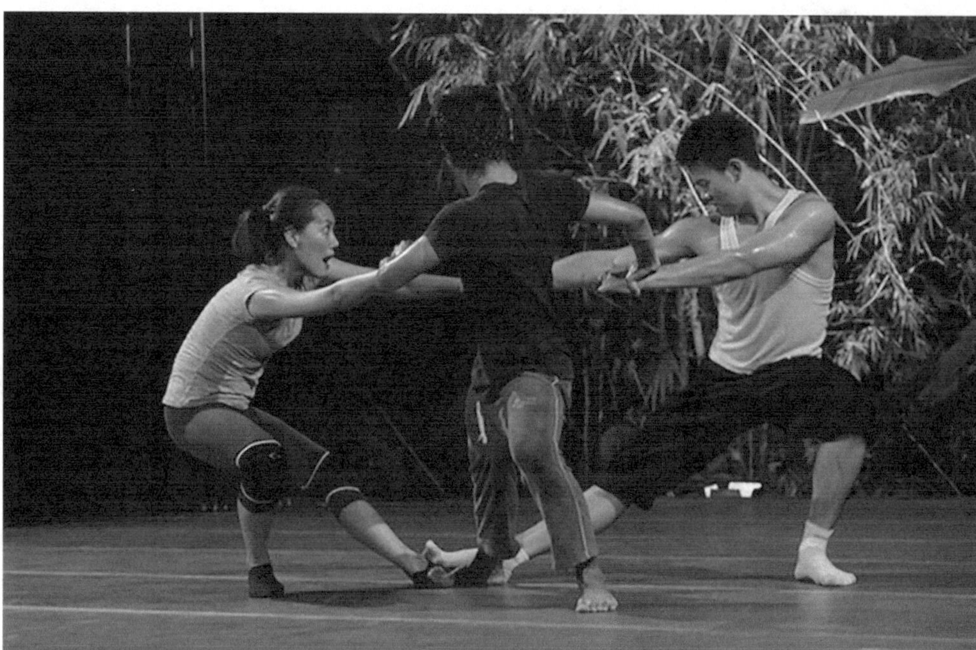

FIGURE 17.4 *Tam Kai* (2013)

Source: Pichet Klunchun Dance Company

Photo: Pichet Klunchun personal collection

Suffering life in relentless motion: *Dancing with Death*

In confronting death over several years since 2014, Thailand faced several turning points.[7] These transitions challenged Pichet to develop a new direction for his Company. To seek new ways to find and exercise freedom for his Company and for Thai dance, Pichet began a project in 2014 that led the Company to delve into new arenas for its dance. It chose to explore the north-eastern Thai district of Dansai in Loei province, where the famous *Phi Ta Khon* ghost festival – like the Hungry Ghost Festival in East Asia – takes place each year in late June or early July.[8] What fascinated Pichet by what he found there was the animated and powerful beauty of the dancers far from the national and familiar centre of Thai culture. These dancers showed unending and fluid *joie de vivre* that suggested an important access point to inspired movement unavailable to artists in Bangkok. The dancers engaged in lively collective performance that spanned dances done in homes to those done on the streets with the spirits amid loud music and free-spirited dance.

In examining the first day of the three-day festival, Pichet saw humans interacting with ghosts in the town. As part of local spirit-mediums inviting the local protector spirit to the town from the nearby river, people become ghost-like by donning colourful costumes and elaborate masks and dancing through local streets to lively Isaan music. To turn this sacred public festival into a dance, Pichet linked it his own experience of daily meditation. By combining his meditation experience and his discoveries in Dansai, Pichet developed a piece which bridged the communal Isaan Festival in Dansai and the Company's desire to experiment more with freedom in Bangkok. By thinking ambitiously, including engaging the dead, Pichet sought to enter into a bigger space for practicing freedom. The result of these efforts was called *Dancing with Death*.

He noted in 2016 to the Esplanade commissioning organization that while waiting for a new king, a new supreme patriarch, and a new government and constitution, Thailand was 'looking for direction yet we are in a state of limbo, uncertain of the near and distant future'.[9] In the midst of these uncertainties, Pichet sought a 'way out of limbo, for my work and society' but as someone who sees these changes as 'an outsider of my own country'.[10] He also sought to move, in the *Dancing with Death* piece, beyond the choreographic safe zone of the Company, which is best known for being rooted in Khon and the *Ramakien*. And to take risks with performance, to move forward, he knew that the desired movement out of the current indeterminate state may not succeed. He sought to engage the folk performance of the *Phi Ta Khon* festival, to be inspired by it and play with what it offered him and his dancers as they explored new paths for moving creatively and embodying intuition. The results were a new corporeal language that sought to give the improvisational and intuitive nature of this unique folk dance a contemporary sensibility. Here, body and mind play off one another in a new in-between zone where material life, connection, and awareness intertwine, seeking to find a way out of the uncertainty.

The focus and the physical centre of the Company's engagement with free movement in *Dancing with Death* was a massive yellow, unevenly twisted oval platform, which he said 'represents the relentless desire or emotion that keeps us moving through cycles of life and death'.[11] The platform is the 'zero' zone which dancers seek to exit, but only outwardly. No one can enter the empty, black centre, where suffering would end. The dominance of the set is a fitting reminder that it was imagined and designed before any choreography for the piece was done. The dancers perform the restless minds and lost souls on the platform,

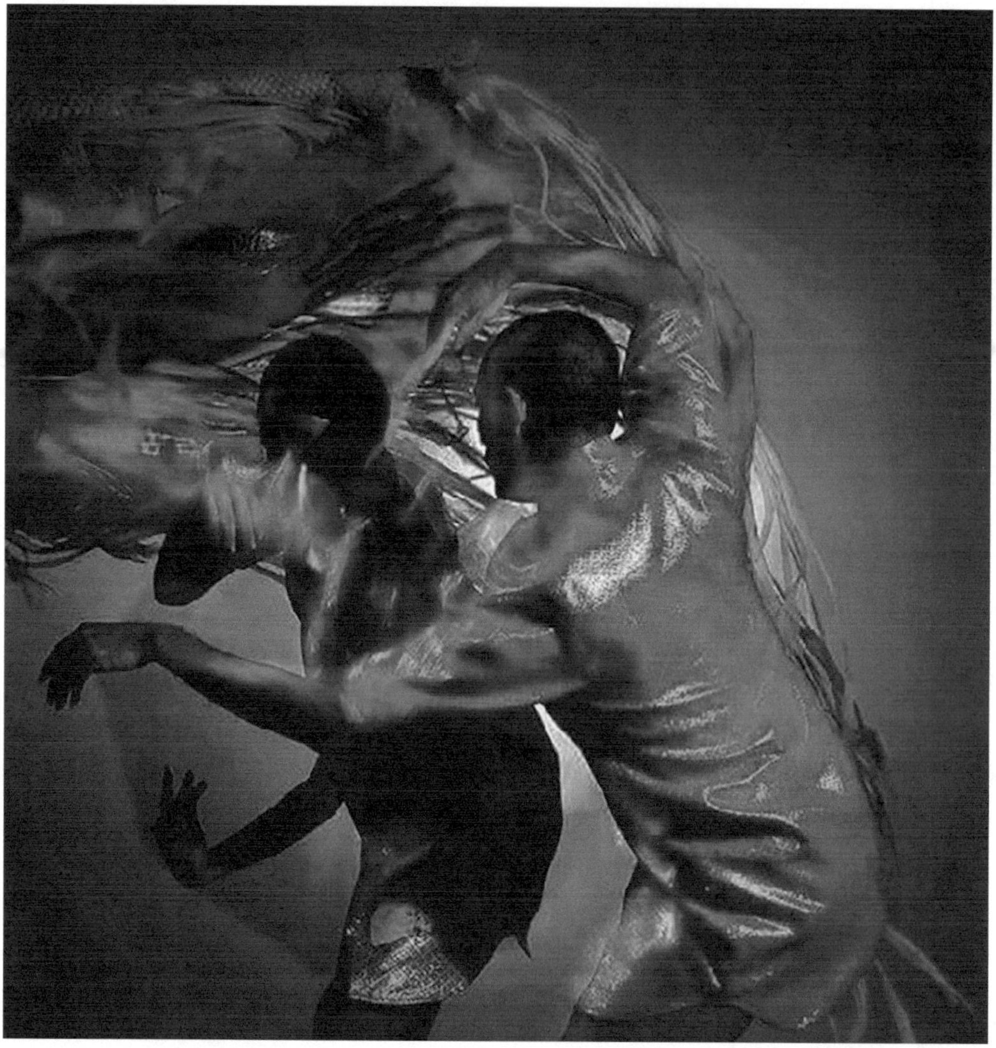

FIGURE 17.5 Pichet Klunchun performs in *Dancing with Death* (2005)

Source: Photo: Pichet Klunchun personal collection

sometimes individually, sometimes in pairs or triplets, and sometimes as an ensemble to changing musical styles.[12] Representing the spiritual remains of humans in a post-mortal limbo world of persistent desire and suffering, its sloping circular path points to the unending circuitous roller coaster of human emotions and experience. To understand these spiritual remains or souls, Pichet was puzzled until he read *Citta is Buddha* by the 19th-century Buddhist monk, Luang Pu Dulaya 'Dun' Atulo of Wat Burapharam in Thailand's Surin province. The English translation by Venerable Bikkhu Khemasanto notes that all Buddhas and all living things share the same *citta* ('soul', often translated as 'mind' but referring to the state of mind or mindset and alluding to the emotive side of the mind), which is intangible, invisible, and indestructible; colourless and formless; indefinable and incomparable.

Mostly performed atop the large yellow platform, *Dancing with Death* shows the relentless motion of the dancers' post-mortal, pre-reincarnated souls. Each of the ensemble's six

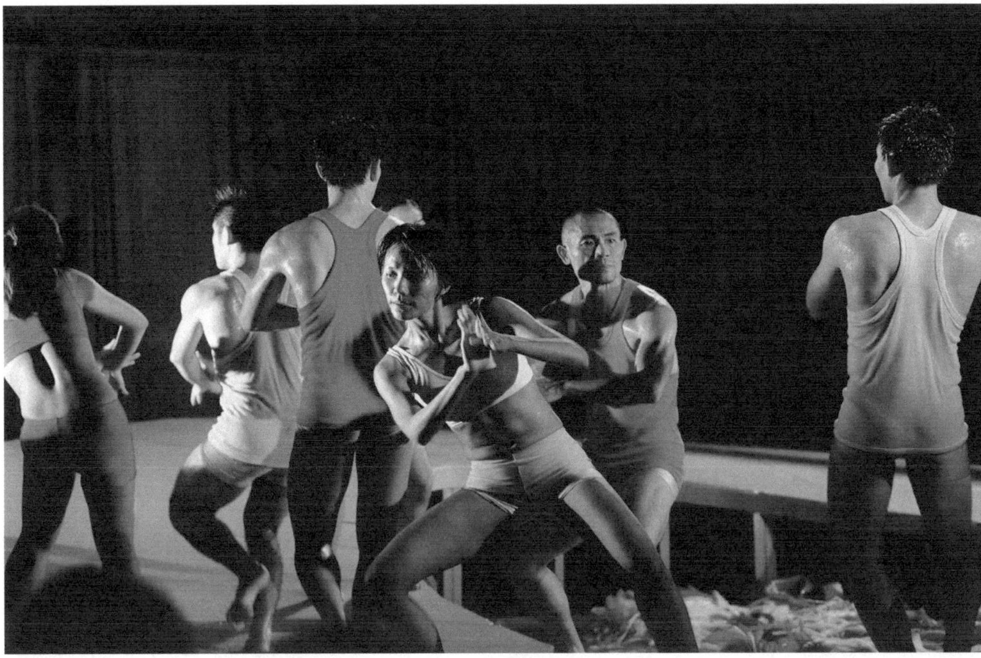

FIGURE 17.6 *Dancing with Death*. Performers try to find their way out of the cycle of suffering in a dance inspired by the Phi Ta Khon festival (2015)

Source: Photo: Pichet Klunchun personal collection

barefoot dancers cycles past and with one another, going round and round the platform, in both directions, sometimes running fast, sometimes walking slow, sometimes stopped, nearly always in motion, often exhausted, in different configurations, but never seeming to find a way off the big yellow treadmill of undead souls they are stuck on. Their footfalls create new rhythmic sounds at odds with the changing music as they crisscross one another on the uneven yellow ghostly surface that is their new home. They are moving through this new shadowy in-between world where their training and their character do not matter. Cycling through relentless patterns of not-quite repeated movement, the dancers develop the piece gradually and subtly, sometimes circling their palms inside out over and over again, giving this simple action a greater force and intensity. This simple, repeated hand motion, a kinetic allusion to the simple intensity of the simple *Phi Ta Khon* performers in Dansai, centres the full motion of the dancers' bodies. These hand rotations are repeated throughout the dance and form one focus for the dancers to improvise. They fall in and out of unison, finding rest in the ordinary where they playfully skip, fervidly crouch, or frenetically jump. Their slow strolls progress to long strides until they sprint over the surface, pausing at the tops of the platform to rest and gather momentum.

After 45 minutes or so of this relentless collective motion, the piece escalates into a hypnotic lull, broken by the distinctive whine of a *kaen* beginning a joyful Isaan song that marks an outward turn of the performance. The performers slowly exit the yellow platform and call for audience members to join them in dancing in the new carnival-like environment that hearkens back to the *Phi Ta Khon* festival in Dansai, Thailand. Some of the performers dress up in the colourful costumes of the wandering ghosts like in Dansai.

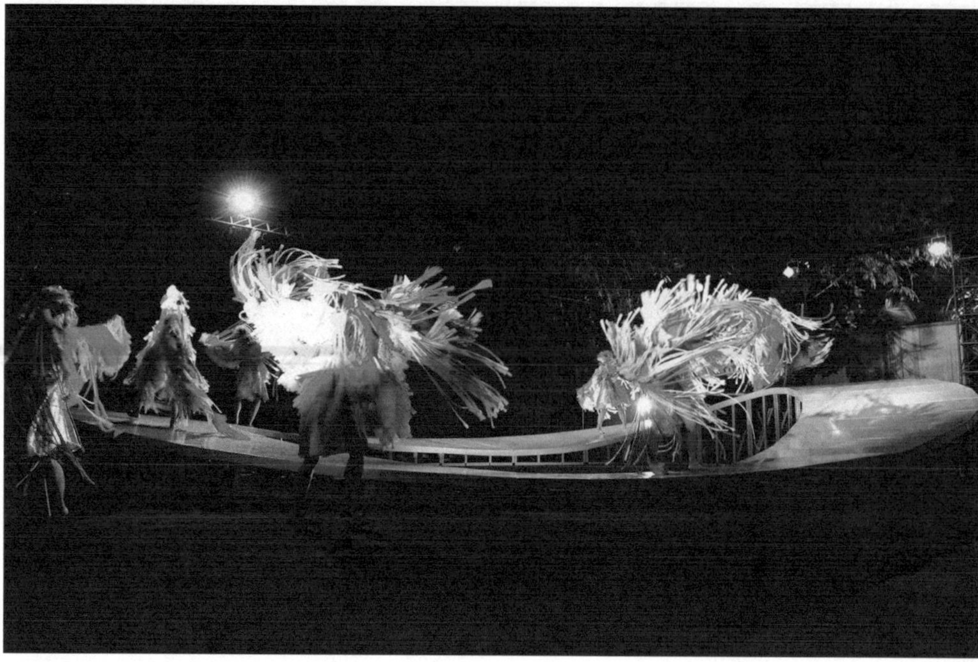

FIGURE 17.7 *Dancing with Death* (2015)

Source: Pichet Klunchun Dance Company

Photo: Pichet Klunchun personal collection

The dancers of the Company return to the performance space and are now dancing with a free vitality off the yellow circuit of post-mortal souls where they had previously been relentlessly cycling, with some from the audience who feel the spirit move them. The costumed dancers, led by Pichet, are now free from the strain of repetition and happy to join the thumping mob, until they move offstage to end the performance.

The *Dancing with Death* experiment did not fulfil its ambitions. Focused on engaging big experiences like death and big concepts like mind and soul, the piece concentrated too much on the dancers' inner conditions and neglected the bodies they needed to do their work. It did not enable the Company's dancers to move on, to freely move on their own terms separate from their training and practice. And although the performance reached four foreign countries between late 2015 and 2017 – first at Pichet's theatre, then in Singapore, Japan, Australia, Taiwan, and again in Thailand – it travelled with difficulty. This difficulty was in no small part due to the great cost of shipping and dis-assembling the great yellow performance platform from venue to venue. By being too ambitious, too big, too inner-focused, too complex, and too costly, the piece was too difficult to move more widely on tour and unable to move the dancers in a new direction. Since that time, the Company has been moving in a lighter, simpler, and more dynamic direction, focusing more on bringing the dancers' bodies back into focus as it searches for ways to bring more freedom into its performances.

The two dance pieces of the Pichet Klunchun Dance Company discussed in this chapter were pivotal in changing the direction and the focus of the Company in the 2010s. *Tam Kai* helped the company to break from its perceived creative dead end by seeking to 'destroy the beautiful forms' of its dancers, as Pichet said. It sought to create a new environment where both the individual and collective spontaneity of dance could emerge. *Dancing with Death* continued from that point, seeking to permit the Company to leave behind the ruined dance forms and to provide opportunities for its dancers to develop modes of embodied free movement different from what they knew or had done in the past. While Pichet has dedicated his career to developing innovative choreographic techniques from Khon, the two pieces discussed here sought to explore dance practice beyond Khon. He experiments extend the possibilities of dance that emerge from the classically trained body while seeking to free the body, mind, and spirit of Thai dance.

The contemporary Thai dance scene

How does the Pichet Klunchun Dance Company fit into Thailand's larger ecology of contemporary dance? Dance is integral to cultural life in Thailand, but there are few dedicated platforms for contemporary dance. There is much graceful, elegant – even beautiful – movement in Thailand today, both on- and offstage, in real-life settings, and onscreen but little conceptual or cultural space for professional contemporary dance. Besides the spontaneous hand, arm, body, face, and leg movements of Thai people in their mundane interactions in markets; at work; on the street; in homes; or through Instagram, YouTube, and TikTok, there are many types of ritualized performance done in temples, shrines, and sacred venues. Beyond these large assemblages of fluid movement patterns visible in daily life, there are many contemporary performing arts groups that make coordinated aesthetic movement onstage key parts of their performances. Ordinary Thais and performing artists seemingly dance freely whenever the music and rhythm move them, even though they may not consider themselves dancers. There are also many government dance schools, private dance schools, university dance departments, and dance events in Thailand, producing many talented dancers and skilled dance groups who perform in a contemporary mode around the country. Dance is clearly an important part of Thai culture.

Yet while a wide range of practiced bodily movements are common in daily life, in ritual, in education, in entertainment, in markets, in traditional and modern stage performances, contemporary dance companies in Thailand are rare; since there is no clearly dedicated platform or cultural setting for contemporary dance as a distinct type of performance activity, it is easy to go unnoticed. Most contemporary dance companies are run by event managers who are looking for dance groups to perform as needed for their events or to train dancers for the high demand for skilful dancers.[13] In such a context, despite this variety of dance in Thailand today, the Pichet Klunchun Dance Company is one of the few professional contemporary *dance* companies based in Thailand known outside of the Kingdom.

Thailand's contemporary dance scene has expanded and improved considerably in the past decade. More and more performers and groups from Thailand are being invited to join international festivals or to work with international artists or groups. This global exposure not only permits those in other countries to see the varied ways that dance is part of contemporary performance in Thailand, but it is also an excellent opportunity for Thai performers and groups to discover how others do their work in the world and to see how their work relates to the work done by artists elsewhere. Thailand's involvement in the world of performance is important to take them beyond what they see in classrooms or performance spaces in the Kingdom. When dance-oriented performers take time to appreciate other art forms, they will be better able to cultivate their work in the future.

Conclusion

The Pichet Klunchun Dance Company stands out for being one of the few professional contemporary dance companies in Thailand known outside the country. Although best known for its reimagining of classical the Khon tradition for the contempory world, its work draws on and incorporates diverse bundles of patterned and spontaneous movement from throughout Thailand into its performances. As such, it is part of a bigger ecology of Thai dance culture. Like other performing groups using dance in Thailand, the Company is part of an evolving and interconnected ecosystem of efforts to produce living movement in Thailand. The two pieces examined in this chapter – *Tam Kai* and *Dancing with Death* – underscore the company's struggle to reinvent itself in the 2010s by destroying the beautiful but familiar embodied forms of its earlier dance and then by seeking new ways for its dancers – and the company as a whole – to discipline themselves to a freer type of Thai dance rooted in their knowledge and experience. Its struggle to reinvent and to practice dance more freely continues within the evolving ecology of contemporary dance in Thailand.[14]

Notes

1 For some rehearsals in Bangkok, since there was not enough room in the studio for developing the project, the Company rehearsed in a nearby covered indoor soccer (futsol) field.
2 For the da:ns festival 2011, Esplanade in Singapore invited Pichet to be the artist-in-residence and commissioned *Black & White*. This opportunity permitted him to form the Pichet Klunchun Dance Company on full-time basis. The production later toured in Belgium, Japan, China, and other places.
3 From interviews with Pichet Klunchun on 1 February and 15 March 2020 (Klunchun 2020). Also see Asia Hundreds (2014).
4 Unstable governments and coups d'état have marked Thailand's history since 1932, including ten coups since 1945. Since 2005, moments of mass protest, military crackdown, and coups continued, in 2005–2006, 2010, and 2013–14, with coups occurring in 2006 and 2014. This 21st-century political unrest has occupied the minds and bodies of many people in Thailand. During this decade, there were often large people's protests which occupied some key parts of Bangkok before being ended by either the military or a coup. Pichet's *Black & White* piece sought to stress the importance of balance, even during apparent conflicts.
5 Interview with Pornrat Damrhung and Lowell Skar, 1 February 2020 (Klunchun 2020).
6 Interview with Pichet on 15 March 2020 (Klunchun 2020).
7 There were several interconnected elements to this transition in Thailand during the 2010s. Besides the 2014 coup and the ensuing military government, there were deaths of major figures in key Thai institutions, especially in the Royal and Buddhist institutions. In 2013, the 100-year-old 19th supreme patriarch of Buddhism in Thailand, Somdet Nyanasamvara Suvaddhana, who had held the position since 1989, died. Due to his frailty, since 2004, Somdet Kiew Upaseṇo had been acting supreme patriarch, but he also died in 2013. This led to the naming in 2013 of the controversial Somdet Chuang Varapuñño to serve as acting supreme patriarch, but he was under investigation for tax evasion, which intensified in the wake of the 2014 coup. After the coup, an 'interim' constitution was put in place in effect in 2014 while a new constitution was drafted and then passed via an August 2016 national referendum before coming into effect in April 2017. More significant, however, was the death, on 13 October 2016, of the beloved King Bhumibol Adulyadej (Rama IX) after a 66-year reign at the age of 88. King Rama IX's heir, Crown Prince Vajiralongkorn, was named King on 1 December 2016 (retroactively dated to 13 October 2016). A new 20th supreme patriarch, Somdet Phra Ariyavongsagatanana IX, was officially named to this position in February 2017 under a new

selection process. Following the cremation of Rama IX on 26 October 2017, King Vajiralongkorn was officially coronated on 4–6 May 2019 (Rama X). These deaths and their transitions during the 2010s contributed to a sense of uncertainty and uneasiness among many Thais during the decade and contributed to the need for Pichet Klunchun's company to change.

8 This piece was commissioned by Esplanade – Theatres on the Bay in Singapore and co-produced by Performing Arts Meeting in Yokohama 2016 Executive Committee, Arts Centre Melbourne – Asia TOPA and Adelaide, and Festival Centre's OzAsia Festival. The *Phi Ta Khon* festival in Dansai is the first day of the three-day Buddhist merit-making celebration called *Bun Pawet* or *Bun Luang* (Asia Hundreds 2014).

9 During 2016, there was much uncertainty, first about the new military government constitution, which was put up for an August referendum; more importantly about what would happen after King Rama IX passed away in October; and finally, about who, if anyone, the new supreme patriarch would be. For the quotation, see Tan (2016), Programme, p. 2.

10 Ibid.

11 Quoted from our interview with him on 15 March 2020.

12 Sending this set to the venues where the piece would be performed was the most challenging part of the project, due to the high costs of delivering it, setting it up, and dismantling it. Besides being performed in Singapore, it also went to Yokohama Japan, Australia, and Taiwan.

13 Pornrat Damrhung and Lowell Skar are currently researching the larger field of contemporary dance in Thailand.

14 Pornrat Damrhung and Lowell Skar are researching more on the ecology of contemporary dance in Thai performance and are on the current phase of work in the Pichet Klunchun Dance Company.

References

Asia Hundreds. 2014. "A Maverick Artist Who Revives Traditional Culture: PICHET KLUNCHUN Interview." Accessed January 15, 2020. https://jfac.jp/en/culture/features/asiahundred02/

Klunchun, Pichet. 2020. Interviews conducted by Lowell Skar and Pornrat Damrhung on February 1 and March 15, 2020.

Tan, Faith. 2016. *Dancing with Death House Programme*. Singapore: Esplanade.

The South Pacific

18

MAKING WAVES

Identity, relationships, and leadership within dance in the South Pacific

Sarah Knox and Rose Martin

Dance within the South Pacific is vibrant and varied (Buck & Rowe, 2014). Heterogeneous dance styles, meanings, and experiences mingle in locations as diverse as dance studios, community halls, car parks, and stages. Dance stories are told and retold as lives shapeshift year to year. The dance ecosystem in the South Pacific becomes more complex as contemporary dance process and performance widens its scope and indigenous dance forms claim space within a historically colonialized landscape (Banks, 2017).

This chapter presents vignetted narratives of three women – Sachiko Soro, Sarah Foster-Sproull, and Cat Ruka – working in dance in the South Pacific. The narratives shared by Sachiko, Sarah, and Cat illuminate their experiences of being mid-career dance practitioners in Fiji and Aotearoa/New Zealand. With intersecting practices and career choices, these three women stand as leaders within the South Pacific dance sector. Here, in this chapter, they communicate their challenges, aspirations, and revelations of searching for an identity in dance, striving for sustainable careers, meaningful relationships, and building community through a desire to give back and create space for those who follow in generations to come.

'Showcasing our stories, the way we want to tell them' – Sachiko Soro

When I started VOU [dance company], I wanted it to be very open, where anyone can create. We're collaborative, and our guiding philosophy is that of a servant leader, which means we serve those who come after us rather than dictating to them. Our organizational structure is an inverted pyramid. I have a strong belief in empowering other people, to make sure that everyone else knows how to do all the work, so that if I dropped dead tomorrow, VOU would be fine. I'm not hoarding power or decision-making. You need to give power away constantly in order for the organization to grow. You have a framework of values that you operate within, but you give freedom. In my opinion, VOU is still a very unique arts organization in Fiji.

One of VOU's founding philosophies is to provide fulfilling and sustainable career pathways for Fijian artists. As a developing country, Fiji does not have government or public

DOI: 10.4324/9781003160007-31

funding for the arts. Because of this, VOU has a very entrepreneurial structure, where we perform art commercially, and that funds our contemporary practice and all of the other things that we want to do. We were originally based in Suva but realized that we needed to diversify if we wanted to support our Fijian artists. This meant moving to Nadi, as there are more opportunities for us. We have set up an amazing performing arts venue, which is the first of its kind on the western side of Fiji.

I sold my family house in order to set up a circus tent, which was the only way we could make it work. We knew that if we wanted this thing to survive economically, the tent needed to be as close as possible to Denarau. Around 80% of the tourists who come to Fiji go to either Denarau or the Coral Coast, and there are 3,300 hotel rooms with an average of 80% occupancy. We found an old sugar cane farm, and we designed an awesome theatre. But the engineer said, 'there's no way you're constructing this theatre, the ground is too soft. You need to put this much soil on it and rest it for a year, and then you can construct the theatre'. I was like, 'no we're not doing that!' So, I started looking online for a circus tent. I found one in Cyprus and it came with everything, the seats, the lights, the sound system. I told my friend, who is an architect, 'I found it, I'm going to get this one'. She said 'you're not going to send 50,000 Euros to random Ukrainian man in Cyprus! You need to see it!' and I thought, 'you have a point'. So, we went to Cyprus and we requested him to put up the tent so she could check the steel. It was beautiful, it seats 450 people, so we bought it. A friend in Latvia packed it into a container and shipped it to Fiji. When it arrived, we did not know how to put it up. I emailed the man who made the tent at the company in Poland, and I asked him if we could fly him to Fiji to teach us. He and his wife are in their 70s, and they came and taught us how to operate the tent. It's quite an art

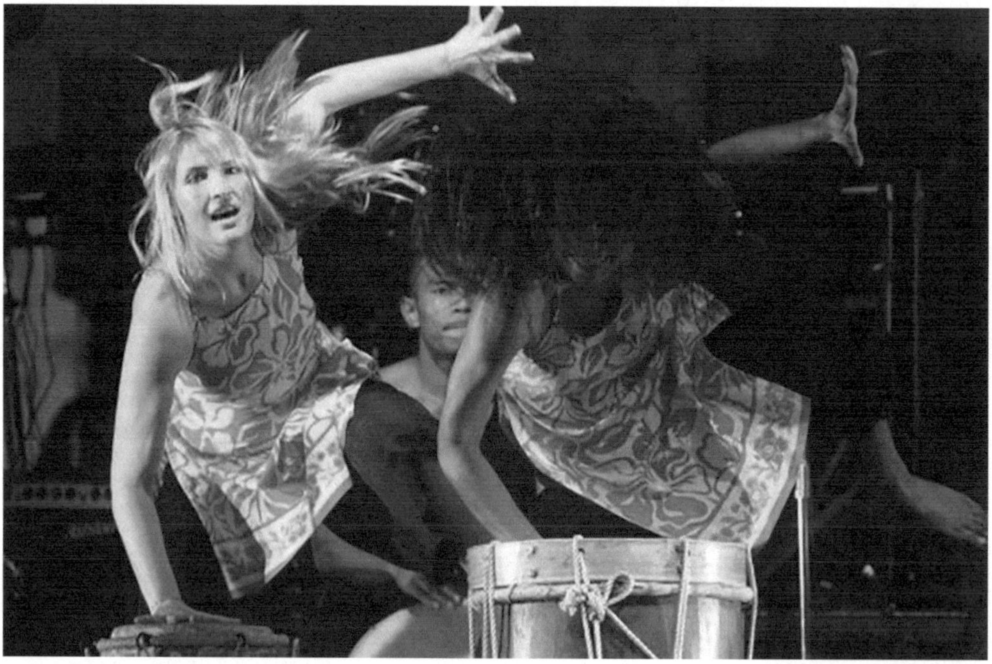

FIGURE 18.1 Sachiko Soro

Source: Photograph: Author's personal collection

form – but now we have our own venue, and we can put on our own show. I am really proud of it because it's made by artists, for artists. We can showcase our stories the way we want to tell them. We share the tent with the local community for free, to get the arts going. I'd love to do festivals there and invite festival buyers from overseas to come, because there are hardly any Pacific arts in the world festival market. It would be really helpful for Pacific artists to be seen on the world stage.

We realized that if the VOU dancers were to have longevity in their careers and to go into different sections of the community to work as artists, we needed a qualification. It needed to be verification that this person has had years of experience as a performing artist; they know repertoire, technique; they've toured; they understand how to make costumes, how to teach and choreograph. That's what prompted us to develop the Tertiary Dance Diploma at the Conservatorium of Dance. It's the first dance qualification in the Pacific Islands. Knowing the graduate profile and what the employment opportunities are in Fiji, we've made it specific to the Fijian context. The curriculum is very practical. We want the dancers to be as versatile in different genres as possible, because we understand what kind of performance might be required in their careers. They all have to sing, play percussion, do traditional Fijian dance, create traditional Fijian costumes. They also do a foundation of classical Indian dance, Bollywood, dances from all around the Pacific – Cook Islands, Samoa, and Tonga. They also do ballet and contemporary technique, hip hop, and jazz. They have to do 40 shows for performing experience. The skills needed to thrive here are a little bit different to in New Zealand. There, there are a lot of employment opportunities in education, in community dance, and in dance research. Those opportunities do not exist in Fiji. In primary and secondary schools in Fiji, there are no artistic subjects such as music, visual art, and dance. So, all of those dance teaching jobs in schools are not possible. There's still so much stigma around dance in Fiji, and people generally don't support their kids to dance. There are a lot of creative people who have dropped out of school, because there is just nothing for them in the curriculum. Because Fiji is a developing country, and most people do not have much money, we had to use a unique educational model. We support the students while they do the course, provide accommodation, give them an allowance; they don't pay fees. Then we guarantee them employment when they finish, and a small percentage of their pay is deducted for the first two years of employment to repay the tuition to keep paying the tutors.

Fifty percent of our working staff at VOU are women. Women in Fiji get a lot of opportunities in the arts. For example, the American embassy has free training for women entrepreneurs, which I've been able to take advantage of. I'm really grateful for all of those sorts of opportunities. We also have some funding to help with marketing, because that was one thing we were really bad at, from the Market Development Facility, which is an Australian government initiative that supports private-sector enterprise. Their main philosophy is to support women into employment, and VOU is doing that. One of the things we do at VOU is to give three months of full paid paternity leave to our male staff. Which might not sound revolutionary, but it is very rare to have three months paid paternity leave in Fiji. I believe this is important because it gives a signal to our men that they are also responsible for childcare, and that burden does not only fall on the woman's shoulders. We have very firm gender roles in Fiji still, and I'm so lucky I have a husband who is supportive, and he looks after the children. I really believe that we need to shift that mindset. Women are generally still expected to cook, to do all of the childcare, and that's why we find women are dropping out of the workforce. When they come back into the workforce, they're not

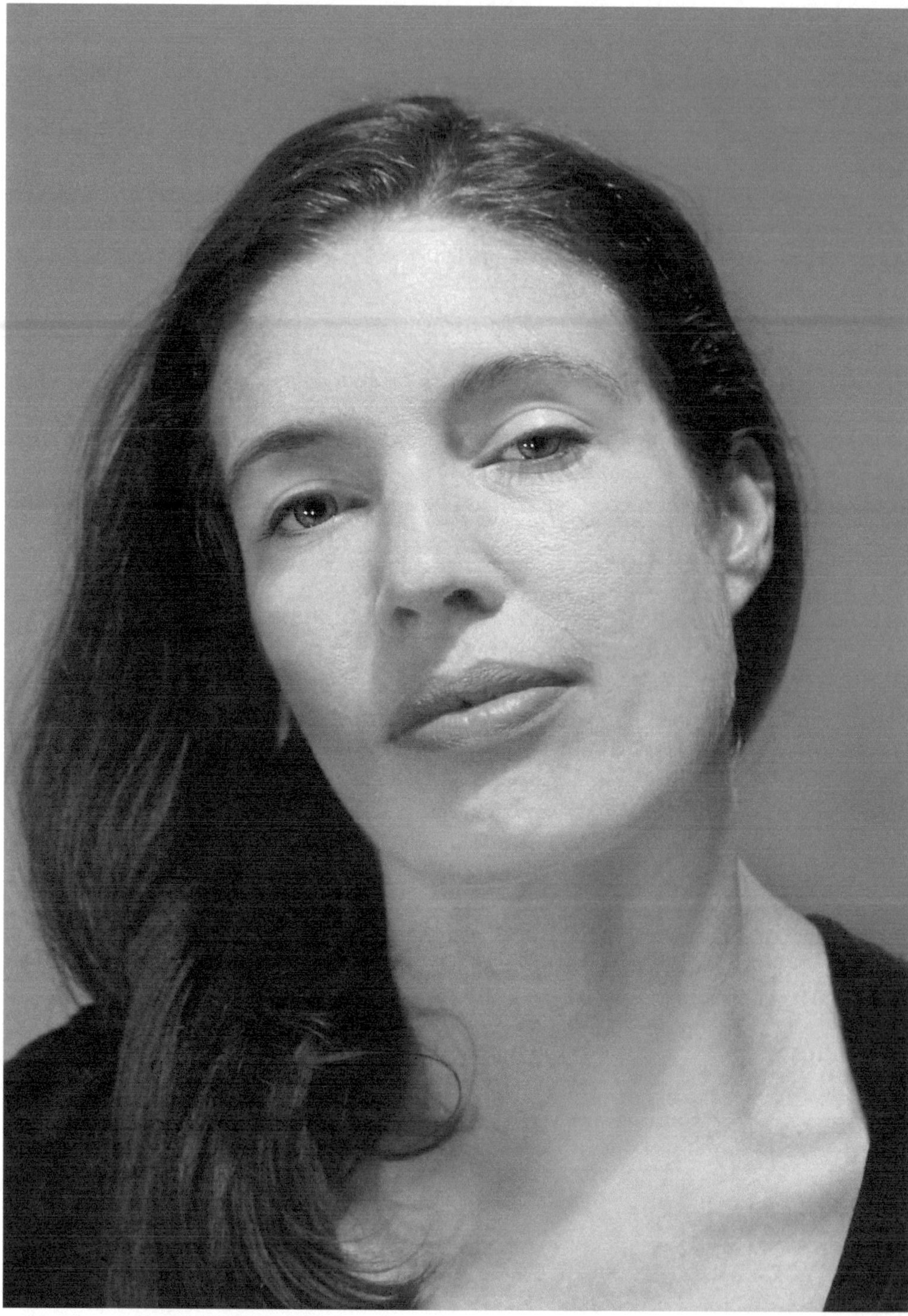

FIGURE 18.2 Sarah Foster-Sproull

Source: Photograph by Andrew Foster

at the same place as they were, whereas men don't have to do that. I'm trying to support women in their relationships and to try to find different ways to keep them employed. For example, if they are dancers, maybe they can do administration work for a while. But most of the dance companies in Fiji are run by women, and all of the women I know in Fiji who are artists are very outspoken!

> **Sachiko Soro**, along with her dance family, VOU, is on a mission: to inspire healing and oneness through the transformative magic of creative experiences. In 2007, Sachiko founded VOU Dance Company in the Fiji Islands. Since its inception, VOU now employs 54 people, mostly Fijian artists. VOU has toured to some of the world's biggest festivals in over 35 countries, there are over 100 students in the VOU Dance School, and VOU has started Fiji's first tertiary dance institution, the Conservatorium of Dance. In 2019, VOU Dance Company opened VOU HUB Fiji, a performance venue that hosts performances by VOU and other artists, for the local and tourist markets. Sachiko is passionate about supporting Fijian artists into fulfilling and sustainable career paths and providing opportunities for young Fijians to research into indigenous Fijian chant and dance. After completing primary and secondary school in Fiji, Sachiko attended the University of Auckland, completing a Bachelor of Music in Composition, a Bachelor of Performing Arts (dance) and a Postgraduate Honours degree (first-class honours) in the Dance Studies Department. Sachiko lives with her dancing machine husband and two energetic sons in Nadi, Fiji.

'Long term relationships in the arts, that's what I'm after' – Sarah Foster-Sproull

I have been a freelance artist for 22 years. Since 2014, I have been focusing on making choreography in earnest and really investing in that process. Foster Group is the dance company I formed that predominantly showcases my choreographic work, and we have an ethos of collaboration and creative facilitation. I really want to invest in relationships with key people over a long period of time, because I understand what it means as an artist to have an ongoing connection with artistic peers. Also, it is challenging to forge a sustainable career in contemporary dance in New Zealand because of the small size of the country. In some ways, I am trying to provide opportunities for my collaborators to relax and know how important they are within my choreographic work by providing them with as much ongoing work as I can. That is my response to recalling how, as a dancer, I felt like I always had to be at the top of my game, that I had to be 'everything' for a choreographer because otherwise the dance work and opportunities might be taken away from me. For me, that generated a tremendous amount of anxiety which took me quite a long time to process after my retirement from dancing. It is only now, after several years, that I actually want to dance again, and I realize that I never actually retired, I just stopped dancing. This is because a traumatic experience in the workplace impacted how I felt about myself as a person, and as a response, I wanted to invest in leading choreographic processes. I love being in charge of the choreographic process, because I am able to make decisions about how the dance making unfolds and what the culture of the working environment is. In its best moment, it is a beautiful collaboration, and we are friends making work together.

The last few years have been a flurry of increased choreographic commissions. I've been tumbling, sort of rolling around, and finding myself in some situations that I never would have imagined. In 2017, I was awarded the Creative New Zealand Choreographic Fellowship, which is a NZ$100,000 grant awarded every two years, for a programme of events that a choreographer designs. I had watched my peers claim that grant in previous years and thought how wonderful it would be to have the resources and time to do something substantial choreographically. I applied and had one of those wonderful phone calls, where Creative New Zealand contacted me to say I had been successful. I was very surprised! Then I got to put into reality some of the things that I had been thinking about, with people I really care about. I was negotiating making work in New Zealand, in Scotland, in China with Footnote New Zealand Dance and Guangdong Modern Dance Company, in Fiji with VOU Dance Company. I did a bunch of creative experiments towards building my picture of a sustainable career as a freelance artist in New Zealand. What I've learned is that sustainability is possible, but long term, it is incredibly difficult. This is because New Zealand has limited dance companies, and I'm predominantly working in a commission context. I feel I can only draw on government resources intermittently. But beyond the fellowship, all of a sudden, I received opportunities that might not have been offered to me previously. This is because the fellowship came with some real mana [prestige] at a time that was very meaningful to me. I'd just had another baby, and completed my master's degree in Dance Studies at The University of Auckland. I was figuring out what life was going to be as an artist mother of two children when I got this huge grant and I could make some creative dreams become real.

Thematically, my career is about nurturing and holding relationships. That's really important to me, that I really invest in my creative relationships deeply, before I obsess too much about my art making. I recognize that some key people in my life have given me opportunities that were well beyond my skill set at that time. For some reason, they saw something in me and took a risk. That is how Patricia Barker (Artistic Director of the Royal New Zealand Ballet) negotiated our relationship initially; her actions toward me said metaphorically, 'I pick you, I'm going to hold the door open for you to come in to this ballet world and I'm going to give you a chance to develop as an artist, and we're going to see where we're going to go with it'. The same investment had come from Deirdre Tarrant (founder Footnote New Zealand Dance). I've had these key female mentors through my artistic life, who held the door open until I could figure out what it is that I am doing. I really appreciate their support, and this is why I want to hold the door open for my dancer-collaborators to come into these same worlds. Long term relationships in the arts, that's what I'm after. I don't want to have any one-night stands; that's not my vibe. Also, I am excited about the potential for arts leadership in Aotearoa. There are brilliant examples of women around me engaging deeply in their sense of what is right in the world. They demonstrate their ethics in the way that they conduct their relationships. I feel quite lucky actually that there are strong examples of women and men, leading with sensibilities towards equality and foregrounding a diverse and inclusive workplace. I'm excited about young artists growing up in an environment where they can see themselves reflected in the people within arts leadership.

I have been involved in Dance Studies at The University of Auckland for a few years, and when a job came up there, I thought, 'That is it! That is the job that I want, that is the one I've been waiting for'. I put my heart and soul into my application and into my interview, and I got the job. I feel very full of gratitude, and while I know that I've worked very hard for it, I also know that it's an absolute privilege. It feels like the culmination of many things converging in one point, because the job at Dance Studies draws together my creative practice research, teaching practice, writing, mentorship and dance community connections. This is what I want to do for the rest of my life. This is the environment I want to work in. I feel it's the next layer of learning; it is the next challenge for my creative life. After 22 years of freelancing, this is my first permanent full-time job. I feel a great responsibility to give back to Dance Studies, to the people that I work with, and in particular to the students. I also want to work harder and undertake even bigger creative challenges. I know that I've been working hard for years, but I recognize that the stability of a full-time job can allow me to work harder at the craft of being a choreographer and a leader and to give more to my community.

Sarah Foster-Sproull is an acclaimed choreographer and dancer working in New Zealand and internationally. She is a Senior Lecturer in Dance Studies at The University of Auckland with a research focus on choreography, creative practice, and contemporary dance technique. Sarah is the Artistic Director of Foster Group Dance and the Choreographer in Residence at the Royal New Zealand Ballet and was Creative New Zealand's Choreographic Fellow for 2017–2019. Sarah has choreographic relationships with Footnote New Zealand Dance, Tamsyn Russell and Dance Base (Edinburgh), Guangdong Modern Dance Company (China), T.H.E. (Singapore), and VOU (Fiji). Her choreographic research traverses large-scale works for up to 100 performers, to intimate performances involving one or two dancers. Sarah was 2019's Director of Choreography at The World of Wearable Art, New Zealand's largest performance event. Sarah is a distinguished graduate of the New Zealand School of Dance and holds a Masters of Dance Studies (first-class Honours) from the University of Auckland.

'Diversity is the key to the world' – Cat Ruka

Since 2019, I have been the artistic director for Tempo Dance Festival, but I started off making choreography and being an artist myself. Through the themes I was exploring in my own work, I fell into teaching – because I was interested in relationships. I became interested in social development and youth organizations not just in dance but across multiple art forms. I became quite heavily involved with rangatahi (young people) and how creativity meets young people. The work that I have done in the lead-up to Tempo has been a massive 'hodge-podge' mix of things to do with guiding others through their creative processes. My interest in working with rangatahi comes from a deep desire to make things better for the next generation, because there were things that I personally really struggled with as a creative young person. I didn't have people from the generation who came before me to guide me in the way that I needed as a young Māori woman. So, connecting with youth is a deeply personal journey for me. There is something super rewarding about being able to help prepare young people for what's ahead but they can't see yet. When I was younger, trying to be a creative person, I thought that there were only one or two ways of being creative that were valued. I didn't understand that what I had to say and that my creative voice were going to be of value. So, I lost my authenticity along the way, because I thought that I had

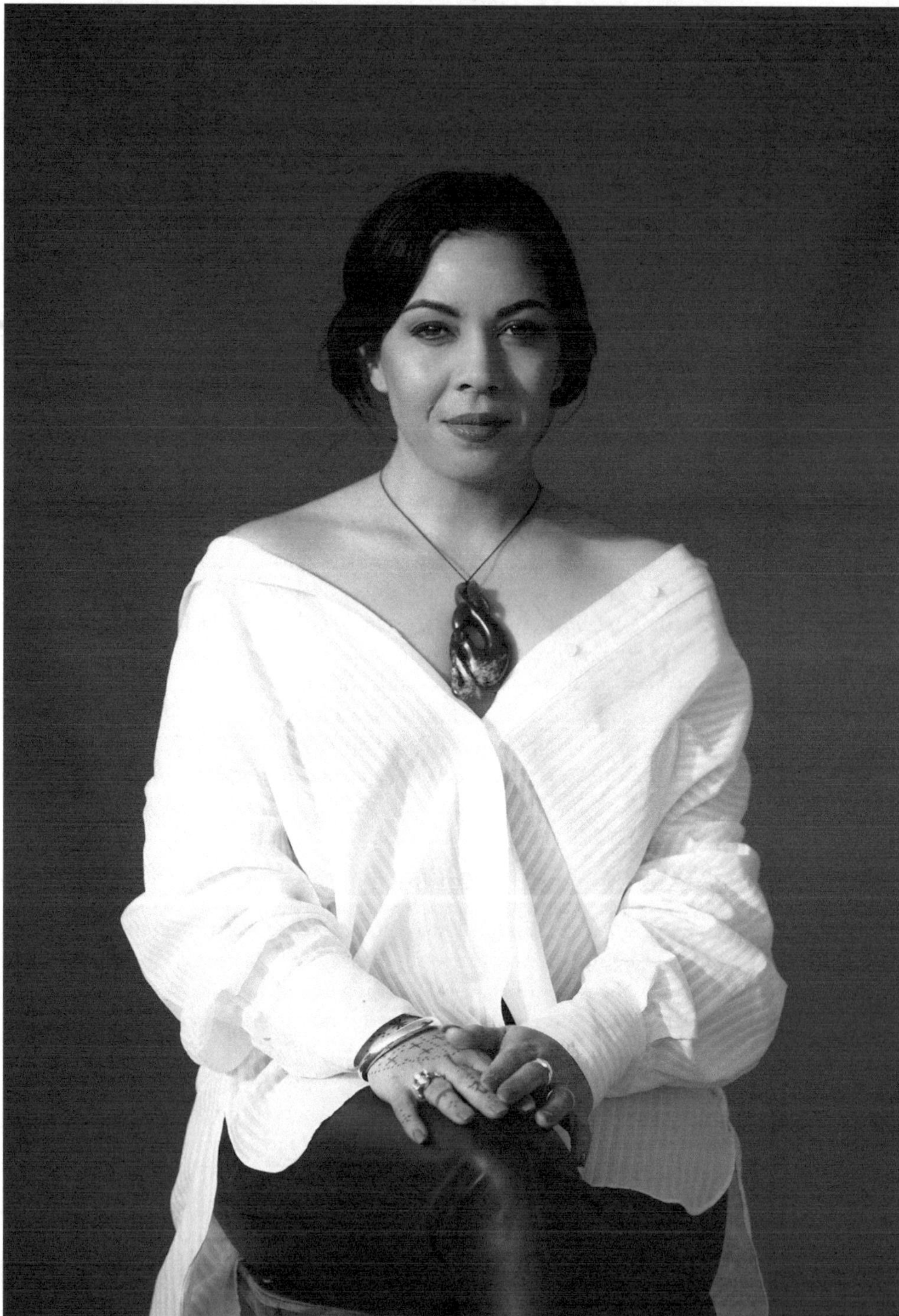

FIGURE 18.3 Cat Ruka

Source: Photograph by Ralph Brown

to do it in the way that was being validated. The work for me is about making it possible for people to stand in their full authentic self as a creative maker. There was this beautiful thing one of my students said once: 'We need to be able to have dangerous conversations in a safe space'. It's about people being able to lay their truths down without feeling as though they're going to be judged. It's about inhabiting the emotion space, of feeling, whether or not that brings up anger, or tears, or elation, joy and gratitude, all of those sorts of things. Whatever it brings up, it means we are all going to hold each other in a way that means no one is ever going to have their integrity compromised by others.

I think the biggest work I'm having to move through currently is what leadership in dance looks like. The word 'leadership' doesn't sit well with me because it's quite an individualistic term. It doesn't really align well with my values. I see myself more as a custodian, or as one of the many people who looks after the dance community. As an arts custodian, you're dealing with feeling, emotion, spirit, and creativity, so there's certain things I do to be an effective leader in the arts. For me, it comes down to four main practices. The first thing is to have a rigorous self-reflection and self-awareness practice. I activate that practice daily. I'm constantly in conversation with my ego and with the ghosts of my past, the trauma I've experienced. I'm constantly in dialogue with my personal desires and separating that from what is actually needed for the community. I think the first pillar of effective arts leadership is self-work. The second pillar I have a practice around is knowing how to connect people together. I think artists need that more than other kinds of people. The Māori word of rangatiratanga directly translates to your ability to weave people together, from lots of different backgrounds and histories. The third thing is strategy. You're essentially responsible for clearing the pathways for the community to inhabit. In order to do that well, you have to have a strategy. You have to understand what's going on in a global arts sphere. You have to have knowledge around what's happening so that you can design a strategy for the people you're responsible for. And the fourth, very personal thing, is around spirituality, having a practice of understanding that there is something much bigger than us. That understanding enables me to not take the work too seriously at the end of the day. Work is small in comparison to things like family, health, and wellness. I try to activate those four practices in order to try to be as effective as possible in my role as a leader.

One of the really special things around Tempo is that it's a pan-genre festival that tries to activate space for lots of different kinds of people and lots of different cultures and a lot of different styles and forms of dance. That really attracted me to the role because it aligns with my personal values, that diversity is the key to the world. Variance and difference are what makes the world beautiful. My main work as the artistic director is to activate that difference of intersectionality, the weaving together of lots of different perspectives of dance. I think in the past, dance has struggled to validate that difference, of form or genre, of style, and of people, so the work really is about validating diversity. I think in New Zealand, we are so blessed and lucky to be connected to all of the Pacific Islands and the huge diversity of dance that comes out of where we are geographically; the work is actually just about lifting the veil on who we are. There's not one answer to who we are. I think the Pacific, Oceania, Te Moana nui a Kiwa is a diaspora, a melting pot of cultures, of ways of viewing the world. I think in Aotearoa, dancers, people working with the body have such an incredible, nuanced offering to make. There are multiple different kinds of beauty. For me, first and foremost, that comes down to our indigenous people, the connection we have to whenua (land), because you can't find that anywhere else in the world.

I always come back to place, whenua, who are our indigenous people? What is the creative realm around our indigenous people? How can we connect to that, and how can that be our anchor strategically moving forward? I think this is key, because this is how we are special. I think our dance institutions are really starting to do that in a meaningful way. We don't just teach western dance forms any more, but when we do, I think we do it in a way that is really connected to who we are and where we are. So, when it comes to strategy for Tempo or for me around leadership, it's about understanding connection to place. In terms of our role as artists in New Zealand, the indigenous people are essentially the spokespeople for our global identity. Artists are often responsible for communicating who we are as people.

Cat Ruka (Ngapuhi, Waitaha) is an acclaimed indigenous arts leader based in Tāmaki Makaurau, Aotearoa. Cat holds a bachelor's degree, post-graduate diploma, and master's degree, all from The University of Auckland's Dance Studies department. She is currently conducting her PhD research in decolonial arts leadership and has presented her research internationally. Trained in choreography, performance art, and indigenous studies, Cat's artistic practice is unique to Aotearoa. Her body of award-winning performance works has toured nationally and internationally, offering relevant discourse into global conversations around contemporary indigenous identity. Her interdisciplinary approach to making work has led her to collaborate with a number of artists from a range of disciplines, including musician Dudley Benson; arts collective Fafswag; designer and stylist Sammy Salsa; photographer Ralph Brown; performance art group COVEN; and a range of independent dance-makers, including experimental choreographers Tru Paraha and Val Smith. Cat has also attended invitational artist residencies in Auckland, Jakarta, New Jersey, New York, and Berlin. In addition to her artistic practice, Cat worked as a performing arts educator and mentor for ten years, predominantly with Māori and Pasifika rangatahi in both mainstream and alternative education settings. She was the Programme Leader at Manukau Institute of Technology's Faculty of Creative Arts and has mentored youth and developed educational programmes for acclaimed creative youth organization Ngā Rangatahi Toa. Cat also does freelance work for not-for-profit arts organizations such as grant writing and creative consultancy and has sat on a number of boards within NZ's arts sector. From 2019–2020, Cat was the Artistic Director of Tempo Dance Festival, New Zealand's premiere dance event that has taken place every year in Tāmaki Makaurau, Auckland, for 20 years. In 2020, Cat has led the organization into a completely digital space in light of the challenges that COVID-19 has created within live performance industries. In 2021, Cat took up the role of Executive Director of Basement Theatre.

References

Banks, O. C. (2017). Haka on the Horizon: Māori contemporary dance and whare tapere. *MAI Journal*, *6*(1), 61–73.

Buck, R. & Rowe, N. (Eds.). (2014). *Moving Oceans: Celebrating Dance in the South Pacific*. New Delhi, India: Routledge.

INDEX

Note: *Italic* page numbers refer to figures.